White Rose Maths

White Rose Maths Edition

Year 1A
A Guide to Teaching for Mastery

Series Editor: Tony Staneff
Lead author: Josh Lury

Pearson

Contents

Introduction to the author team

Power Maths arises from the work of maths mastery experts who are committed to proving that, given the right mastery mindset and approach, **everyone can do maths**. Based on robust research and best practice from around the world, *Power Maths* was developed in partnership with a group of UK teachers to make sure that it not only meets our children's wide-ranging needs but also aligns with the National Curriculum in England.

Power Maths – White Rose Maths edition

This edition of *Power Maths* has been developed and updated by:

Tony Staneff, Series Editor and Author

Vice Principal at Trinity Academy, Halifax, Tony also leads a team of mastery experts who help schools across the UK to develop teaching for mastery via nationally recognised CPD courses, problem-solving and reasoning resources, schemes of work, assessment materials and other tools.

Josh Lury, Lead Author

Josh is a specialist maths teacher, author and maths consultant with a passion for innovative and effective maths education.

The first edition of *Power Maths* was developed by a team of experienced authors, including:

- **Tony Staneff and Josh Lury**

- **Trinity Academy Halifax** (Michael Gosling CEO, Emily Fox, Kate Henshall, Rebecca Holland, Stephanie Kirk, Stephen Monaghan and Rachel Webster)

- **David Board, Belle Cottingham, Jonathan East, Tim Handley, Derek Huby, Neil Jarrett, Stephen Monaghan, Beth Smith, Tim Weal, Paul Wrangles** – skilled maths teachers and mastery experts

- **Cherri Moseley** – a maths author, former teacher and professional development provider

- **Professors Liu Jian and Zhang Dan**, Series Consultants and authors, and their team of mastery expert authors: **Wei Huinv, Huang Lihua, Zhu Dejiang, Zhu Yuhong, Hou Huiying, Yin Lili, Zhang Jing, Zhou Da and Liu Qimeng**

   Used by over 20 million children, Professor Liu Jian's textbook programme is one of the most popular in China. He and his author team are highly experienced in intelligent practice and in embedding key maths concepts using a C-P-A approach.

- **A group of 15 teachers and maths co-ordinators**

   We consulted our teacher group throughout the development of *Power Maths* to ensure we are meeting their real needs in the classroom.

What is *Power Maths*?

Created especially for UK primary schools, and aligned with the new National Curriculum, *Power Maths* is a whole-class, textbook-based mastery resource that empowers every child to understand and succeed. *Power Maths* rejects the notion that some people simply 'can't do' maths. Instead, it develops growth mindsets and encourages hard work, practice and a willingness to see mistakes as learning tools.

Best practice consistently shows that mastery of small, cumulative steps builds a solid foundation of deep mathematical understanding. *Power Maths* combines interactive teaching tools, high-quality textbooks and continuing professional development (CPD) to help you equip children with a deep and long-lasting understanding. Based on extensive evidence, and developed in partnership with practising teachers, *Power Maths* ensures that it meets the needs of children in the UK.

Power Maths and Mastery

Power Maths makes mastery practical and achievable by providing the structures, pathways, content, tools and support you need to make it happen in your classroom.

To develop mastery in maths, children must be enabled to acquire a deep understanding of maths concepts, structures and procedures, step by step. Complex mathematical concepts are built on simpler conceptual components and when children understand every step in the learning sequence, maths becomes transparent and makes logical sense. Interactive lessons establish deep understanding in small steps, as well as effortless fluency in key facts such as tables and number bonds. The whole class works on the same content and no child is left behind.

Power Maths

⚡ Builds every concept in small, progressive steps

⚡ Is built with interactive, whole-class teaching in mind

⚡ Provides the tools you need to develop growth mindsets

⚡ Helps you check understanding and ensure that every child is keeping up

⚡ Establishes core elements such as intelligent practice and reflection

The *Power Maths* approach

Everyone can!

Founded on the conviction that every child can achieve, *Power Maths* enables children to build number fluency, confidence and understanding, step by step.

Child-centred learning

Children master concepts one step at a time in lessons that embrace a concrete-pictorial-abstract (C-P-A) approach, avoid overload, build on prior learning and help them see patterns and connections. Same-day intervention ensures sustained progress.

Continuing professional development

Embedded teacher support and development offer every teacher the opportunity to continually improve their subject knowledge and manage whole-class teaching for mastery.

Whole-class teaching

An interactive, whole-class teaching model encourages thinking and precise mathematical language and allows children to deepen their understanding as far as they can.

What's different in the new edition?

If you have previously used the first editions of *Power Maths*, you might be interested to know how this edition is different. All of the improvements described below are based on feedback from *Power Maths* customers.

Changes to units and the progression

⚡ The order of units has been slightly adjusted, creating closer alignment between adjacent year groups, which will be useful for mixed age teaching.

⚡ The flow of lessons has been improved within units to optimise the pace of the progression and build in more recap where needed. For key topics, the sequence of lessons gives more opportunities to build up a solid base of understanding. Other units have fewer lessons than before, where appropriate, making it possible to fit in all the content.

⚡ Overall, the lessons put more focus on the most essential content for that year, with less time given to non-statutory content.

⚡ The progression of lessons matches the steps in the new White Rose Maths schemes of learning.

Lesson resources

⚡ There is a Quick recap for each lesson in the Teacher Guide, which offers an alternative lesson starter to the Power Up for cases where you feel it would be more beneficial to surface prerequisite learning than general number fluency.

⚡ In the **Discover** and **Share** sections there is now more of a progression from 1 a) to 1 b). Whereas before, 1 b) was mainly designed as a separate question, now 1 a) leads directly into 1 b). This means that there is an improved whole-class flow, and also an opportunity to focus on the logic and skills in more detail. As a teacher, you will be using 1 a) to lead the class into the thinking, then 1 b) to mould that thinking into the core new learning of the lesson.

⚡ In the **Share** section, for KS1 in particular, the number of different models and representations has been reduced, to support the clarity of thinking prompted by the flow from 1 a) into 1 b).

⚡ More fluency questions have been built into the guided and independent practice.

⚡ Pupil pages are as easy as possible for children to access independently. The pages are less full where this supports greater focus on key ideas and instructions. Also, more freedom is offered around answer format, with fewer boxes scaffolding children's responses; squared paper backgrounds are used in the Practice Books where appropriate. Artwork has also been revisited to ensure the highest standards of accessibility.

New components

480 Individual Practice Games are available in *ActiveLearn* for practising key facts and skills in Years 1 to 6. These are designed in an arcade style, to feel like fun games that children would choose to play outside school. They can be accessed via the Pupil World for homework or additional practice in school – and children can earn rewards. There are Support, Core and Extend levels to allocate, with Activity Reporting available for the teacher. There is a Quick Guide on *ActiveLearn* and you can use the Help area for support in setting up child accounts.

There is also a new set of lesson video resources on the Professional Development tile, designed for in-school training in 10- to 20-minute bursts. For each part of the *Power Maths* lesson sequence, there is a slide deck with embedded video, which will facilitate discussions about how you can take your *Power Maths* teaching to the next level.

Your *Power Maths* resources

Pupil Textbooks

Discover, **Share** and **Think together** sections promote discussion and introduce mathematical ideas logically, so that children understand more easily.

Using a Concrete-Pictorial-Abstract approach, clear mathematical models help children to make connections and grasp concepts.

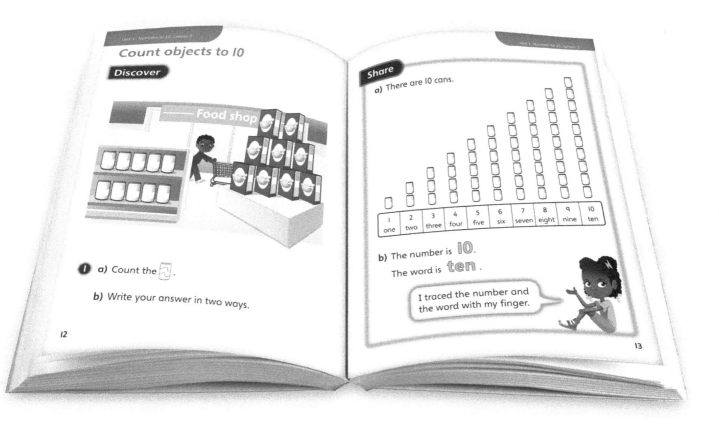

Appealing scenarios stimulate curiosity, helping children to identify the maths problem and discover patterns and relationships for themselves.

Friendly, supportive characters help children develop a growth mindset by prompting them to think, reason and reflect.

To help you teach for mastery, *Power Maths* comprises a variety of high-quality resources.

The coherent *Power Maths* lesson structure carries through into the vibrant, high-quality textbooks. Setting out the core learning objectives for each class, the lesson structure follows a carefully mapped journey through the curriculum and supports children on their journey to deeper understanding.

Pupil Practice Books

The Practice Books offer just the right amount of intelligent practice for children to complete independently in the final section of each lesson.

Practice questions are finely tuned to move children forward in their thinking and to reveal misconceptions.

The practice questions are for everyone – each question varies one small element to move children on in their thinking.

Calculations are connected so that children think about the underlying concept.

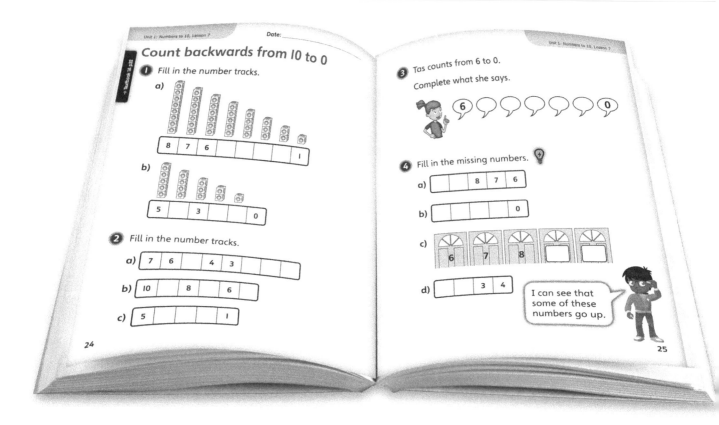

Challenge questions allow children to delve deeper into a concept.

The *Power Maths* characters support and encourage children to think and work in different ways.

Think differently questions encourage children to use reasoning as well as their mathematical knowledge to reach a solution.

Reflect questions reveal the depth of each child's understanding before they move on.

Online subscription

The online subscription will give you access to additional resources and answers from the Textbook and Practice Book.

eTextbooks

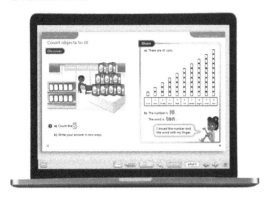

Digital versions of *Power Maths* Textbooks allow class groups to share and discuss questions, solutions and strategies. They allow you to project key structures and representations at the front of the class, to ensure all children are focusing on the same concept.

Teaching tools

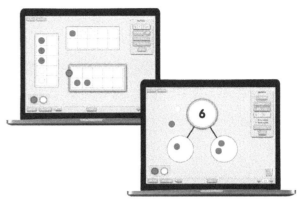

Here you will find interactive versions of key *Power Maths* structures and representations.

Power Ups

Use this series of daily activities to promote and check number fluency.

Online versions of Teacher Guide pages

PDF pages give support at both unit and lesson levels. You will also find help with key strategies and templates for tracking progress.

Unit videos

Watch the professional development videos at the start of each unit to help you teach with confidence. The videos explore common misconceptions in the unit, and include intervention suggestions as well as suggestions on what to look out for when assessing mastery in your students.

End of unit Strengthen and Deepen materials

The Strengthen activity at the end of every unit addresses a key misconception and can be used to support children who need it. The Deepen activities are designed to be low ceiling/high threshold and will challenge those children who can understand more deeply. These resources will help you ensure that every child understands and will help you keep the class moving forward together. These printable activities provide an optional resource bank for use after the assessment stage.

Individual Practice Games

These enjoyable games can be used at home or at school to embed key number skills (see page 6).

Professional Development videos and slides

These slides and videos of *Power Maths* lessons can be used for ongoing training in short bursts or to support new staff.

The *Power Maths* teaching model

At the heart of *Power Maths* is a clearly structured teaching and learning process that helps you make certain that every child masters each maths concept securely and deeply. For each year group, the curriculum is broken down into core concepts, taught in units. A unit divides into smaller learning steps – lessons. Step by step, strong foundations of cumulative knowledge and understanding are built.

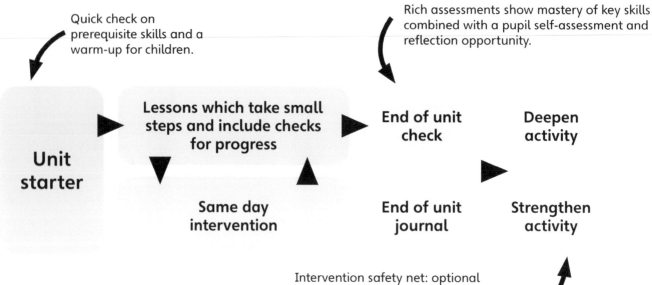

Quick check on prerequisite skills and a warm-up for children.

Rich assessments show mastery of key skills combined with a pupil self-assessment and reflection opportunity.

Unit starter → Lessons which take small steps and include checks for progress → End of unit check → Deepen activity

Same day intervention

End of unit journal

Strengthen activity

Intervention safety net: optional activities to use if assessment shows some children still have misconceptions.

Unit starter

Each unit begins with a unit starter, which introduces the learning context along with key mathematical vocabulary and structures and representations.

- The Textbooks include a check on readiness and a warm-up task for children to complete.

- Your Teacher Guide gives support right from the start on important structures and representations, mathematical language, common misconceptions and intervention strategies.

- Unit-specific videos develop your subject knowledge and insights so you feel confident and fully equipped to teach each new unit. These are available via the online subscription.

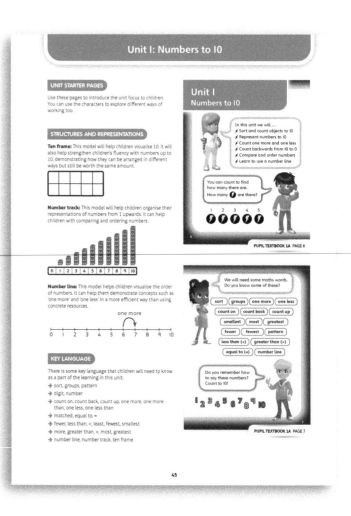

Lesson

Once a unit has been introduced, it is time to start teaching the series of lessons.

- Each lesson is scaffolded with Textbook and Practice Book activities and begins with a Power Up activity (available via online subscription) or the Quick recap activity in the Teacher Guide (see page 15).

- *Power Maths* identifies lesson by lesson what concepts are to be taught.

- Your Teacher Guide offers lots of support for you to get the most from every child in every lesson. As well as highlighting key points, tricky areas and how to handle them, you will also find question prompts to check on understanding and clarification on why particular activities and questions are used.

Same-day intervention

Same-day interventions are vital in order to keep the class progressing together. This can be during the lesson as well as afterwards (see page 27). Therefore, *Power Maths* provides plenty of support throughout the journey.

- Intervention is focused on keeping up now, not catching up later, so interventions should happen as soon as they are needed.

- Practice section questions are designed to bring misconceptions to the surface, allowing you to identify these easily as you circulate during independent practice time.

- Child-friendly assessment questions in the Teacher Guide help you identify easily which children need to strengthen their understanding.

End of unit check and journal

For each unit, the End of unit check in the Textbook lets you see which children have mastered the key concepts, which children have not and where their misconceptions lie. The Practice Books also include an End of unit journal in which children can reflect on what they have learned. Each unit also offers Strengthen and Deepen activities, available via the online subscription.

The Teacher Guide offers different ways of managing the End of unit assessments as well as giving support with handling misconceptions.

The End of unit check presents multiple-choice questions. Children think about their answer, decide on a solution and explain their choice.

The End of unit journal is an opportunity for children to test out their learning and reflect on how they feel about it. Tackling the 'journal' problem reveals whether a child understands the concept deeply enough to move on to the next unit.

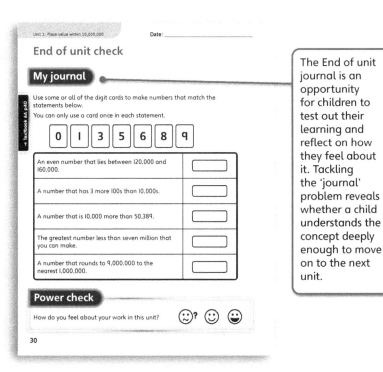

The *Power Maths* lesson sequence

At the heart of *Power Maths* is a unique lesson sequence designed to empower children to understand core concepts and grow in confidence. Embracing the National Centre for Excellence in the Teaching of Mathematics' (NCETM's) definition of mastery, the sequence guides and shapes every *Power Maths* lesson you teach.

Flexibility is built into the *Power Maths* programme so there is no one-to-one mapping of lessons and concepts and you can pace your teaching according to your class. While some children will need to spend longer on a particular concept (through interventions or additional lessons), others will reach deeper levels of understanding. However, it is important that the class moves forward together through the termly schedules.

Power Up ⏱ 5 minutes

Each lesson begins with a Power Up activity (available via the online subscription) which supports fluency in key number facts.

The whole-class approach depends on fluency, so the Power Up is a powerful and essential activity.

The Quick recap is an alternative starter, for when you think some or all children would benefit more from revisiting pre-requisite work (see page 15).

TOP TIP
If the class is struggling with the task, revisit it later and check understanding.

Power Ups reinforce the two key things that are essential for success: times-tables and number bonds.

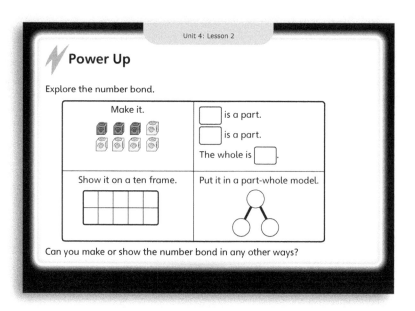

Discover ⏱ 10 minutes

A practical, real-life problem arouses curiosity. Children find the maths through story telling.

TOP TIP
Discover works best when run at tables, in pairs with concrete objects.

Question ❶ a) tackles the key concept and question ❶ b) digs a little deeper. Children have time to explore, play and discuss possible strategies.

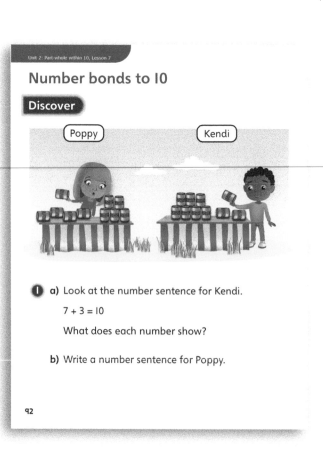

Unit 2: Part-whole within 10, Lesson 7

Number bonds to 10

Discover

❶ a) Look at the number sentence for Kendi.

$7 + 3 = 10$

What does each number show?

b) Write a number sentence for Poppy.

92

Share ⏱ 10 minutes

Teacher-led, this interactive section follows the **Discover** activity and highlights the variety of methods that can be used to solve a single problem.

TOP TIP
You can use the carpet area if you have this. Pairs sharing a textbook is a great format for **Share**!

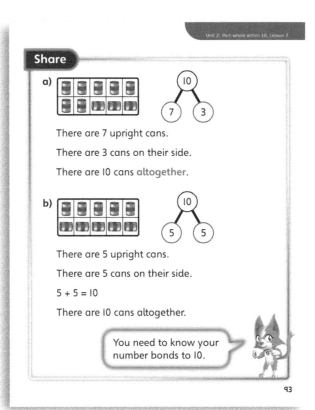

Share

a) There are 7 upright cans.

There are 3 cans on their side.

There are 10 cans altogether.

b) There are 5 upright cans.

There are 5 cans on their side.

$5 + 5 = 10$

There are 10 cans altogether.

You need to know your number bonds to 10.

93

Your Teacher Guide gives target questions for children. The online toolkit provides interactive structures and representations to link concrete and pictorial to abstract concepts.

Bring children to the front to share and celebrate their solutions and strategies.

Think together

⏱ 10 minutes

Children work in groups on the carpet or at tables, using their textbooks or eBooks.

TOP TIP
Make sure children have mini whiteboards or pads to write on if they are not at their tables.

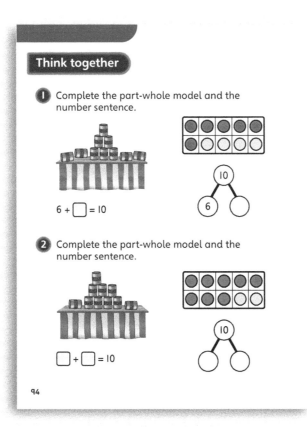

Think together

1 Complete the part-whole model and the number sentence.

$6 + \square = 10$

2 Complete the part-whole model and the number sentence.

$\square + \square = 10$

94

Using the Teacher Guide, model question 1 for your class.

Question 2 is less structured. Children will need to think together in their groups, then discuss their methods and solutions as a class.

Question 3 – the openness of the **Challenge** question helps to check depth of understanding.

13

Using their Practice Books, children work independently while you circulate and check on progress.

Questions follow small steps of progression to deepen learning.

Some children could work separately with a teacher or assistant.

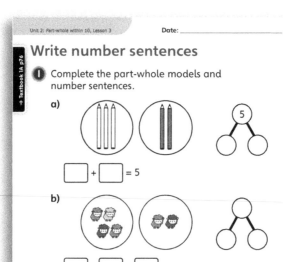

Unit 2: Part-whole within 10, Lesson 3   Date: _____

→ Textbook 1A p76

Write number sentences

① Complete the part-whole models and number sentences.

a)

☐ + ☐ = 5

b)

☐ + ☐ = ☐

② Write a number sentence to match the part-whole model.

7
6   1

☐ + ☐ = ☐

56

Are some children struggling? If so, work with them as a group, using mathematical structures and representations to support understanding as necessary.

There are no set routines: for real understanding, children need to think about the problem in different ways.

'Spot the mistake' questions are great for checking misconceptions.

The **Reflect** section is your opportunity to check how deeply children understand the target concept.

Unit 2: Part-whole within 10, Lesson 3

⑤ Take 6 counters.   CHALLENGE

Make two groups from the counters.

⬤ ⬤ ⬤ ⬤ ⬤ ⬤

Write a number sentence to show the groups that you have made.

_____

_____

Reflect

Write your own number sentence.

Talk about it with a partner.

58

The Practice Books use various approaches to check that children have fully understood each concept.

Looking like they understand is not enough! It is essential that children can show they have grasped the concept.

Using the *Power Maths* Teacher Guide

Think of your Teacher Guides as *Power Maths* handbooks that will guide, support and inspire your day-to-day teaching. Clear and concise, and illustrated with helpful examples, your Teacher Guides will help you make the best possible use of every individual lesson. They also provide wrap-around professional development, enhancing your own subject knowledge and helping you to grow in confidence about moving your children forward together.

There is a Teacher Guide per year group for every term, with unit and lesson level guidance and support.

Never feel stuck! You will find ideas for introducing every unit and lesson and questions to encourage teacher reflection before and after each lesson.

Tips and advice on key elements such as C-P-A approaches, misconceptions, language, modelling growth mindsets and same day intervention.

Annotations for every Textbook and Practice Book page, providing prompts for key questions to ask to expose understanding and explanations as to why key questions have been chosen.

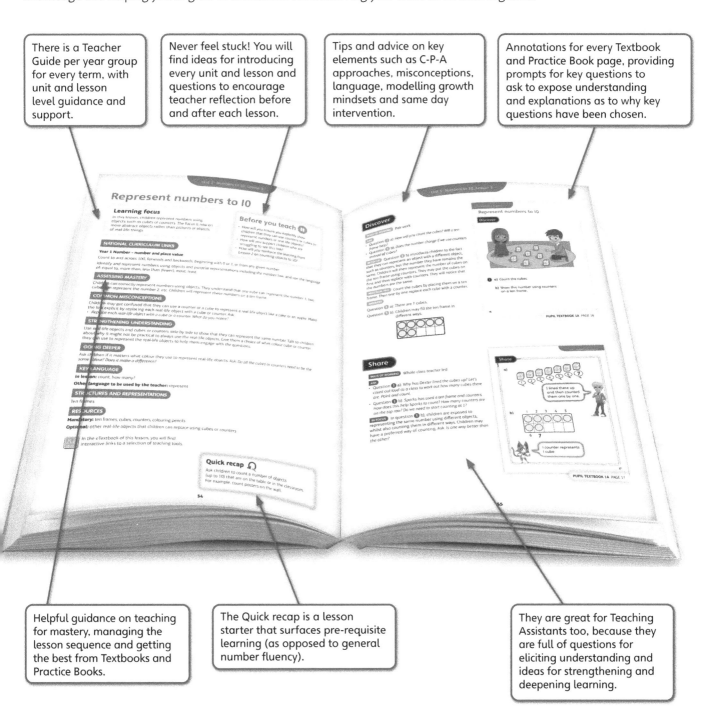

Helpful guidance on teaching for mastery, managing the lesson sequence and getting the best from Textbooks and Practice Books.

The Quick recap is a lesson starter that surfaces pre-requisite learning (as opposed to general number fluency).

They are great for Teaching Assistants too, because they are full of questions for eliciting understanding and ideas for strengthening and deepening learning.

At the end of each unit, your Teacher Guide helps you identify who has fully grasped the concept, who has not and how to move every child forward. This is covered later in the Assessment strategies section.

Power Maths Year I, yearly overview

Textbook	Strand	Unit		Number of lessons
Textbook A / Practice Book A	Number – number and place value	1	Numbers to 10	14
	Number – addition and subtraction	2	Part-whole within 10	7
(Term 1)	Number – addition and subtraction	3	Addition awithin 10	4
	Number – addition and subtraction	4	Subtraction within 10	8
	Geometry – properties of shape	5	2D and 3D shapes	5
Textbook B / Practice Book B	Number – number and place value	6	Numbers to 20	12
	Number – addition and subtraction	7	Addition and subtraction within 20	11
(Term 2)	Number – number and place value	8	Numbers to 50	7
	Measurement	9	Introducing length and height	4
	Measurement	10	Introducing weight and volume	7
Textbook C / Practice Book C	Number – multiplication and division	11	Multiplication and division	9
	Number – fractions	12	Halves and quarters	4
(Term 3)	Geometry – position and direction	13	Position and direction	5
	Number – number and place value	14	Numbers to 100	6
	Measurement	15	Money	3
	Measurement	16	Time	5

Power Maths Year I, Textbook IA (Term I) overview

Strand	Unit		Lesson number	Lesson title	NC Objective 1	NC Objective 2
Number – number and place value	Unit 1	Numbers to 10	1	Sort objects	Identify and represent numbers using objects and pictorial representations including the number line, and use the language of: equal to, more than, less than (fewer), most, least	
Number – number and place value	Unit 1	Numbers to 10	2	Count objects to 10	Count to and across 100, forwards and backwards, beginning with 0 or 1, or from any given number	Identify and represent numbers using objects and pictorial representations including the number line, and use the language of: equal to, more than, less than (fewer), most, least
Number – number and place value	Unit 1	Numbers to 10	3	Represent numbers to 10	Identify and represent numbers using objects and pictorial representations including the number line, and use the language of: equal to, more than, less than (fewer), most, least	Identify and represent numbers using objects and pictorial representations including the number line, and use the language of: equal to, more than, less than (fewer), most, least
Number – number and place value	Unit 1	Numbers to 10	4	Count objects from a larger group	Count to and across 100, forwards and backwards, beginning with 0 or 1, or from any given number	Identify and represent numbers using objects and pictorial representations including the number line, and use the language of: equal to, more than, less than (fewer), most, least
Number – number and place value	Unit 1	Numbers to 10	5	Count on from any number	Count to and across 100, forwards and backwards, beginning with 0 or 1, or from any given number	Identify and represent numbers using objects and pictorial representations including the number line, and use the language of: equal to, more than, less than (fewer), most, least
Number – number and place value	Unit 1	Numbers to 10	6	One more	Given a number, identify one more and one less	Identify and represent numbers using objects and pictorial representations including the number line, and use the language of: equal to, more than, less than (fewer), most, least
Number – number and place value	Unit 1	Numbers to 10	7	Count backwards from 10 to 0	Count to and across 100, forwards and backwards, beginning with 0 or 1, or from any given number	
Number – number and place value	Unit 1	Numbers to 10	8	One less	Given a number, identify one more and one less	
Number – number and place value	Unit 1	Numbers to 10	9	Compare groups	Identify and represent numbers using objects and pictorial representations including the number line, and use the language of: equal to, more than, less than (fewer), most, least	
Number – number and place value	Unit 1	Numbers to 10	10	Fewer or more?	Identify and represent numbers using objects and pictorial representations including the number line, and use the language of: equal to, more than, less than (fewer), most, least	
Number – number and place value	Unit 1	Numbers to 10	11	<, > or =	Identify and represent numbers using objects and pictorial representations including the number line, and use the language of: equal to, more than, less than (fewer), most, least	
Number – number and place value	Unit 1	Numbers to 10	12	Compare numbers	Identify and represent numbers using objects and pictorial representations including the number line, and use the language of: equal to, more than, less than (fewer), most, least	

Strand	Unit		Lesson number	Lesson title	NC Objective 1	NC Objective 2
Number – number and place value	Unit 1	Numbers to 10	13	Order objects and numbers	Identify and represent numbers using objects and pictorial representations including the number line, and use the language of: equal to, more than, less than (fewer), most, least	
Number – number and place value	Unit 1	Numbers to 10	14	The number line	Identify and represent numbers using objects and pictorial representations including the number line, and use the language of: equal to, more than, less than (fewer), most, least	
Number – addition and subtraction	Unit 2	Part-whole within 10	1	Parts and wholes	Identify and represent numbers using objects and pictorial representations including the number line, and use the language of: equal to, more than, less than (fewer), most, least	Represent and use number bonds and related subtraction facts within 20
Number – addition and subtraction	Unit 2	Part-whole within 10	2	The part-whole model	Represent and use number bonds and related subtraction facts within 20	
Number – addition and subtraction	Unit 2	Part-whole within 10	3	Write number sentences	Read, write and interpret mathematical statements involving addition (+), subtraction (–) and equals (=) signs	Represent and use number bonds and related subtraction facts within 20
Number – addition and subtraction	Unit 2	Part-whole within 10	4	Fact families – addition facts	Read, write and interpret mathematical statements involving addition (+), subtraction (–) and equals (=) signs	Represent and use number bonds and related subtraction facts within 20
Number – addition and subtraction	Unit 2	Part-whole within 10	5	Number bonds	Represent and use number bonds and related subtraction facts within 20	
Number – addition and subtraction	Unit 2	Part-whole within 10	6	Find number bonds	Represent and use number bonds and related subtraction facts within 20	
Number – addition and subtraction	Unit 2	Part-whole within 10	7	Number bonds to 10	Represent and use number bonds and related subtraction facts within 20	
Number – addition and subtraction	Unit 3	Addition within 10	1	Add together	Represent and use number bonds and related subtraction facts within 20	
Number – addition and subtraction	Unit 3	Addition within 10	2	Add more	Represent and use number bonds and related subtraction facts within 20	
Number – addition and subtraction	Unit 3	Addition within 10	3	Addition problems	Solve one-step problems that involve addition and subtraction, using concrete objects and pictorial representations, and missing number problems such as 7 = – 9	
Number – addition and subtraction	Unit 3	Addition within 10	4	Find the missing number	Represent and use number bonds and related subtraction facts within 20	
Number – addition and subtraction	Unit 4	Subtraction within 10	1	How many are left? (1)	Represent and use number bonds and related subtraction facts within 20	
Number – addition and subtraction	Unit 4	Fractions (1)	2	How many are left? (2)	Represent and use number bonds and related subtraction facts within 20	
Number – addition and subtraction	Unit 4	Fractions (1)	3	Break apart (1)	Represent and use number bonds and related subtraction facts within 20	
Number – addition and subtraction	Unit 4	Fractions (1)	4	Break apart (2)	Represent and use number bonds and related subtraction facts within 20	

Strand	Unit		Lesson number	Lesson title	NC Objective 1	NC Objective 2
Number – addition and subtraction	Unit 4	Fractions (1)	5	Fact families	Represent and use number bonds and related subtraction facts within 20	
Number – addition and subtraction	Unit 4	Fractions (1)	6	Subtraction on a number line	Solve one-step problems that involve addition and subtraction, using concrete objects and pictorial representations, and missing number problems such as 7 = [ ] – 9	
Number – addition and subtraction	Unit 4	Fractions (1)	7	Add or subtract 1 or 2	Add and subtract one-digit and two-digit numbers to 20, including zero	
Number – addition and subtraction	Unit 4	Fractions (1)	8	Solve word problems – addition and subtraction	Solve one-step problems that involve addition and subtraction, using concrete objects and pictorial representations, and missing number problems such as 7 = [ ] – 9	
Geometry – properties of shape	Unit 5	2D and 3D shapes	1	Recognise and name 3D shapes	Recognise and name common 2D and 3D shapes, including: 3D shapes [for example, cuboids (including cubes), pyramids and spheres]	
Geometry – properties of shape	Unit 5	2D and 3D Shapes	2	Sort 3D shapes	Recognise and name common 2D and 3D shapes, including: 3D shapes [for example, cuboids (including cubes), pyramids and spheres]	
Geometry – properties of shape	Unit 5	2D and 3D Shapes	3	Recognise and name 2D shapes	Recognise and name common 2D and 3D shapes, including: 3D shapes [for example, cuboids (including cubes), pyramids and spheres]	
Geometry – properties of shape	Unit 5	2D and 3D Shapes	4	Sort 2D shapes	Recognise and name common 2D and 3D shapes, including: 3D shapes [for example, cuboids (including cubes), pyramids and spheres]	
Geometry – properties of shape	Unit 5	2D and 3D Shapes	5	Make patterns with shapes	Recognise and name common 2D and 3D shapes, including: 3D shapes [for example, cuboids (including cubes), pyramids and spheres]	Non-statutory guidance: They recognise and create repeating patterns with objects and with shapes

Mindset: an introduction

Global research and best practice deliver the same message: learning is greatly affected by what learners perceive they can or cannot do. What is more, it is also shaped by what their parents, carers and teachers perceive they can do. Mindset – the thinking that determines our beliefs and behaviours – therefore has a fundamental impact on teaching and learning.

Everyone can!

Power Maths and mastery methods focus on the distinction between 'fixed' and 'growth' mindsets (Dweck, 2007).[1] Those with a fixed mindset believe that their basic qualities (for example, intelligence, talent and ability to learn) are pre-wired or fixed: 'If you have a talent for maths, you will succeed at it. If not, too bad!' By contrast, those with a growth mindset believe that hard work, effort and commitment drive success and that 'smart' is not something you are or are not, but something you become. In short, everyone can do maths!

Key mindset strategies

A growth mindset needs to be actively nurtured and developed. *Power Maths* offers some key strategies for fostering healthy growth mindsets in your classroom.

It is okay to get it wrong

Mistakes are valuable opportunities to re-think and understand more deeply. Learning is richer when children and teachers alike focus on spotting and sharing mistakes as well as solutions.

Praise hard work

Praise is a great motivator, and by focusing on praising effort and learning rather than success, children will be more willing to try harder, take risks and persist for longer.

Mind your language!

The language we use around learners has a profound effect on their mindsets. Make a habit of using growth phrases, such as, 'Everyone can!', 'Mistakes can help you learn' and 'Just try for a little longer'. The king of them all is one little word, 'yet'... I can't solve this...yet!' Encourage parents and carers to use the right language too.

Build in opportunities for success

The step-by-small-step approach enables children to enjoy the experience of success. In addition, avoid ability grouping and encourage every child to answer questions and explain or demonstrate their methods to others.

[1]Dweck, C (2007) *The New Psychology of Success*, Ballantine Books: New York

The *Power Maths* characters

The *Power Maths* characters model the traits of growth mindset learners and encourage resilience by prompting and questioning children as they work. Appearing frequently in the Textbooks and Practice Books, they are your allies in teaching and discussion, helping to model methods, alternatives and misconceptions, and to pose questions. They encourage and support your children, too: they are all hardworking, enthusiastic and unafraid of making and talking about mistakes.

Meet the team!

Creative Flo is open-minded and sometimes indecisive. She likes to think differently and come up with a variety of methods or ideas.

Determined Dexter is resolute, resilient and systematic. He concentrates hard, always tries his best and he'll never give up – even though he doesn't always choose the most efficient methods!

'Let's try again.'

'Mistakes are cool!'

'Have I found all of the solutions?'

'Let's try it this way…'

'Can we do it differently?'

'I've got another way of doing this!'

'I'm going to try this!'

'I know how to do that!'

'Want to share my ideas?'

Curious Ash is eager, interested and inquisitive, and he loves solving puzzles and problems. Ash asks lots of questions but sometimes gets distracted.

'What if we tried this…?'

'I wonder…'

'Is there a pattern here?'

Sparks the Cat

Miaow!

Brave Astrid is confident, willing to take risks and unafraid of failure. She's never scared to jump straight into a problem or question, and although she often makes simple mistakes, she's happy to talk them through with others.

Mathematical language

Traditionally, we in the UK have tended to try simplifying mathematical language to make it easier for young children to understand. By contrast, evidence and experience show that by diluting the correct language, we actually mask concepts and meanings for children. We then wonder why they are confused by new and different terminology later down the line! *Power Maths* is not afraid of 'hard' words and avoids placing any barriers between children and their understanding of mathematical concepts. As a result, we need to be deliberate, precise and thorough in building every child's understanding of the language of maths. Throughout the Teacher Guides you will find support and guidance on how to deliver this, as well as individual explanations throughout the pupil Textbooks.

Use the following key strategies to build children's mathematical vocabulary, understanding and confidence.

Precise and consistent

Everyone in the classroom should use the correct mathematical terms in full, every time. For example, refer to 'equal parts', not 'parts'. Used consistently, precise maths language will be a familiar and non-threatening part of children's everyday experience.

Full sentences

Teachers and children alike need to use full sentences to explain or respond. When children use complete sentences, it both reveals their understanding and embeds their knowledge.

Stem sentences

These important sentences help children express mathematical concepts accurately, and are used throughout the *Power Maths* books. Encourage children to repeat them frequently, whether working independently or with others. Examples of stem sentences are:

'4 is a part, 5 is a part, 9 is the whole.'

'There are .... groups. There are .... in each group.'

Key vocabulary

The unit starters highlight essential vocabulary for every lesson. In the pupil books, characters flag new terminology and the Teacher Guide lists important mathematical language for every unit and lesson. New terms are never introduced without a clear explanation.

Symbolic language

Symbols are used early on so that children quickly become familiar with them and their meaning. Often, the *Power Maths* characters will highlight the connection between language and particular symbols.

The role of talk and discussion

When children learn to talk purposefully together about maths, barriers of fear and anxiety are broken down and they grow in confidence, skills and understanding. Building a healthy culture of 'maths talk' empowers their learning from day one.

Explanation and discussion are integral to the *Power Maths* structure, so by simply following the books, your lessons will stimulate structured talk. The following key 'maths talk' strategies will help you strengthen that culture and ensure that every child is included.

Sentences, not words

Encourage children to use full sentences when reasoning, explaining or discussing maths. This helps both speaker and listeners to clarify their own understanding. It also reveals whether or not the speaker truly understands, enabling you to address misconceptions as they arise.

Working together

Working with others in pairs, groups or as a whole class is a great way to support maths talk and discussion. Use different group structures to add variety and challenge. For example, children could take timed turns for talking, work independently alongside a 'discussion buddy', or perhaps play different *Power Maths* character roles within their group.

Think first – then talk

Provide clear opportunities within each lesson for children to think and reflect, so that their talk is purposeful, relevant and focused.

Give every child a voice

Where the 'hands up' model allows only the more confident child to shine, *Power Maths* involves everyone. Make sure that no child dominates and that even the shyest child is encouraged to contribute – and praised when they do.

Assessment strategies

Teaching for mastery demands that you are confident about what each child knows and where their misconceptions lie; therefore, practical and effective assessment is vitally important.

Formative assessment within lessons

The **Think together** section will often reveal any confusions or insecurities; try ironing these out by doing the first **Think together** question as a class. For children who continue to struggle, you or your Teaching Assistant should provide support and enable them to move on.

▶ Performance in practice can be very revealing: check Practice Books and listen out both during and after practice to identify misconceptions.

▶ The **Reflect** section is designed to check on the all-important depth of understanding. Be sure to review how the children performed in this final stage before you teach the next lesson.

End of unit check – Textbook

Each unit concludes with a summative check to help you assess quickly and clearly each child's understanding, fluency, reasoning and problem solving skills. Your Teacher Guide will suggest ideal ways of organising a given activity and offer advice and commentary on what children's responses mean. For example, 'What misconception does this reveal?'; 'How can you reinforce this particular concept?'

For Year 1 and Year 2 children, assess in small, teacher-led groups, giving each child time to think and respond while also consolidating correct mathematical language. Assessment with young children should always be an enjoyable activity, so avoid one-to-one individual assessments, which they may find threatening or scary. If you prefer, the End of unit check can be carried out as a whole-class group using whiteboards and Practice Books.

End of unit check – Practice Book

The Practice Book contains further opportunities for assessment, and can be completed by children independently whilst you are carrying out diagnostic assessment with small groups. Your Teacher Guide will advise you on what to do if children struggle to articulate an explanation – or perhaps encourage you to write down something they have explained well. It will also offer insights into children's answers and their implications for next learning steps. It is split into three main sections, outlined below.

My journal is designed to allow children to show their depth of understanding of the unit. It can also serve as a way of checking that children have grasped key mathematical vocabulary. The question children should answer is first presented in the Textbook in the Think! section. This provides an opportunity for you to discuss the question first as a class to ensure children have understood their task. Children should have some time to think about how they want to answer the question, and you could ask them to talk to a partner about their ideas. Then children should write their answer in their Practice Book, using the word bank provided to help them with vocabulary.

The **Power check** allows pupils to self-assess their level of confidence on the topic by colouring in different smiley faces. You may want to introduce the faces as follows:

Each unit ends with either a Power play or a Power puzzle. This is an activity, puzzle or game that allows children to use their new knowledge in a fun, informal way.

Progress Tests

There are *Power Maths* Progress Tests for each half term and at the end of the year, including an Arithmetic test and Reasoning test in each case. You can enter results in the online markbook to track and analyse results and see the average for all schools' results. The tests use a 6-step scale to show results against age-related expectation.

How to ask diagnostic questions

The diagnostic questions provided in children's Practice Books are carefully structured to identify both understanding and misconceptions (if children answer in a particular way, you will know why). The simple procedure below may be helpful:

Ask the question, offering the selection of answers provided.

Children take time to think about their response.

Each child selects an answer and shares their reasoning with the group.

Give minimal and neutral feedback (for example, 'That's interesting', or 'Okay').

Ask, 'Why did you choose that answer?', then offer an opportunity to change their mind by providing one correct and one incorrect answer.

Note which children responded and reasoned correctly first time and everyone's final choices.

Reflect that together, we can get the right answer.

Keeping the class together

Traditionally, children who learn quickly have been accelerated through the curriculum. As a consequence, their learning may be superficial and will lack the many benefits of enabling children to learn with and from each other.

By contrast, *Power Maths'* mastery approach values real understanding and richer, deeper learning above speed. It sees all children learning the same concept in small, cumulative steps, each finding and mastering challenge at their own level. Remember that when you teach for mastery, EVERYONE can do maths! Those who grasp a concept easily have time to explore and understand that concept at a deeper level. The whole class therefore moves through the curriculum at broadly the same pace via individual learning journeys.

For some teachers, the idea that a whole class can move forward together is revolutionary and challenging. However, the evidence of global good practice clearly shows that this approach drives engagement, confidence, motivation and success for all learners, and not just the high flyers. The strategies below will help you keep your class together on their maths journey.

Mix it up

Do not stick to set groups at each table. Every child should be working on the same concept, and mixing up the groupings widens children's opportunities for exploring, discussing and sharing their understanding with others.

Recycling questions

Reuse the Textbook and Practice Book questions with concrete materials to allow children to explore concepts and relationships and deepen their understanding. This strategy is especially useful for reinforcing learning in same-day interventions.

Strengthen at every opportunity

The next lesson in a *Power Maths* sequence always revises and builds on the previous step to help embed learning. These activities provide golden opportunities for individual children to strengthen their learning with the support of Teaching Assistants.

Prepare to be surprised!

Children may grasp a concept quickly or more slowly. The 'fast graspers' won't always be the same individuals, nor does the speed at which a child understands a concept predict their success in maths. Are they struggling or just working more slowly?

Same-day intervention

Since maths competence depends on mastering concepts one by one in a logical progression, it is important that no gaps in understanding are ever left unfilled. Same-day interventions – either within or after a lesson – are a crucial safety net for any child who has not fully made the small step covered that day. In other words, intervention is always about keeping up, not catching up, so that every child has the skills and understanding they need to tackle the next lesson. That means presenting the same problems used in the lesson, with a variety of concrete materials to help children model their solutions.

We offer two intervention strategies below, but you should feel free to choose others if they work better for your class.

Within-lesson intervention

The **Think together** activity will reveal those who are struggling, so when it is time for practice, bring these children together to work with you on the first practice questions. Observe these children carefully, ask questions, encourage them to use concrete models and check that they reach and can demonstrate their understanding.

After-lesson intervention

You might like to use the **Think together** questions to recap the lesson with children who are working behind expectations during assembly time. Teaching Assistants could also work with these children at other convenient points in the school day. Some children may benefit from revisiting work from the same topic in the previous year group. Note also the suggestion for recycling questions from the Textbook and Practice Book with concrete materials on page 26.

The role of practice

Practice plays a pivotal role in the *Power Maths* approach. It takes place in class groups, smaller groups, pairs, and independently, so that children always have the opportunities for thinking as well as the models and support they need to practise meaningfully and with understanding.

Intelligent practice

In *Power Maths*, practice never equates to the simple repetition of a process. Instead we embrace the concept of intelligent practice, in which all children become fluent in maths through varied, frequent and thoughtful practice that deepens and embeds conceptual understanding in a logical, planned sequence. To see the difference, take a look at the following examples.

Traditional practice

- Repetition can be rote – no need for a child to think hard about what they are doing

- Praise may be misplaced

- Does this prove understanding?

Intelligent practice

- Varied methods – concrete, pictorial and abstract

- Equation expressed in different ways, requiring thought and understanding

- Constructive feedback

All practice questions are designed to move children on and reveal misconceptions.

Simple, logical steps build onto earlier learning.

C-P-A runs throughout – different ways of modelling and understanding the same concept.

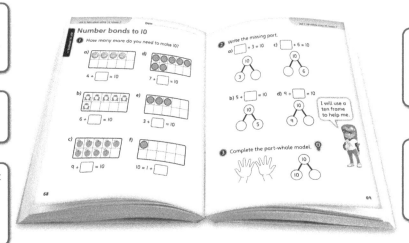

Conceptual variation – children work on different representations of the same maths concept.

Friendly characters offer support and encourage children to try different approaches.

A carefully designed progression

The Practice Books provide just the right amount of intelligent practice for children to complete independently in the final sections of each lesson. It is really important that all children are exposed to the practice questions, and that children are not directed to complete different sections. That is because each question is different and has been designed to challenge children to think about the maths they are doing. The questions become more challenging so children grasping concepts more quickly will start to slow down as they progress. Meanwhile, you have the chance to circulate and spot any misconceptions before they become barriers to further learning.

Homework and the role of parents and carers

While *Power Maths* does not prescribe any particular homework structure, we acknowledge the potential value of practice at home. For example, practising fluency in key facts, such as number bonds and times-tables, is an ideal homework task. You can share the Individual Practice Games for homework (see pages 6 and 9), or parents and carers could work through uncompleted Practice Book questions with children at either primary stage.

However, it is important to recognise that many parents and carers may themselves lack confidence in maths, and few, if any, will be familiar with mastery methods. A Parents' and Carers' evening that helps them understand the basics of mindsets, mastery and mathematical language is a great way to ensure that children benefit from their homework. It could be a fun opportunity for children to teach their families that everyone can do maths!

Structures and representations

Unlike most other subjects, maths comprises a wide array of abstract concepts – and that is why children and adults so often find it difficult. By taking a concrete-pictorial-abstract (C-P-A) approach, *Power Maths* allows children to tackle concepts in a tangible and more comfortable way.

Non-linear stages

Concrete

Replacing the traditional approach of a teacher working through a problem in front of the class, the concrete stage introduces real objects that children can use to 'do' the maths – any familiar object that a child can manipulate and move to help bring the maths to life. It is important to appreciate, however, that children must always understand the link between models and the objects they represent. For example, children need to first understand that three cakes could be represented by three pretend cakes, and then by three counters or bricks. Frequent practice helps consolidate this essential insight. Although they can be used at any time, good concrete models are an essential first step in understanding.

Pictorial

This stage uses pictorial representations of objects to let children 'see' what particular maths problems look like. It helps them make connections between the concrete and pictorial representations and the abstract maths concept. Children can also create or view a pictorial representation together, enabling discussion and comparisons. The *Power Maths* teaching tools are fantastic for this learning stage, and bar modelling is invaluable for problem solving throughout the primary curriculum.

Abstract

Our ultimate goal is for children to understand abstract mathematical concepts, symbols and notation and of course, some children will reach this stage far more quickly than others. To work with abstract concepts, a child must be comfortable with the meaning of and relationships between concrete, pictorial and abstract models and representations. The C-P-A approach is not linear, and children may need different types of models at different times. However, when a child demonstrates with concrete models and pictorial representations that they have grasped a concept, we can be confident that they are ready to explore or model it with abstract symbols such as numbers and notation.

Use at any time and with any age to support understanding

Variation helps visualisation

Children find it much easier to visualise and grasp concepts if they see them presented in a number of ways, so be prepared to offer and encourage many different representations.

For example, the number six could be represented in various ways:

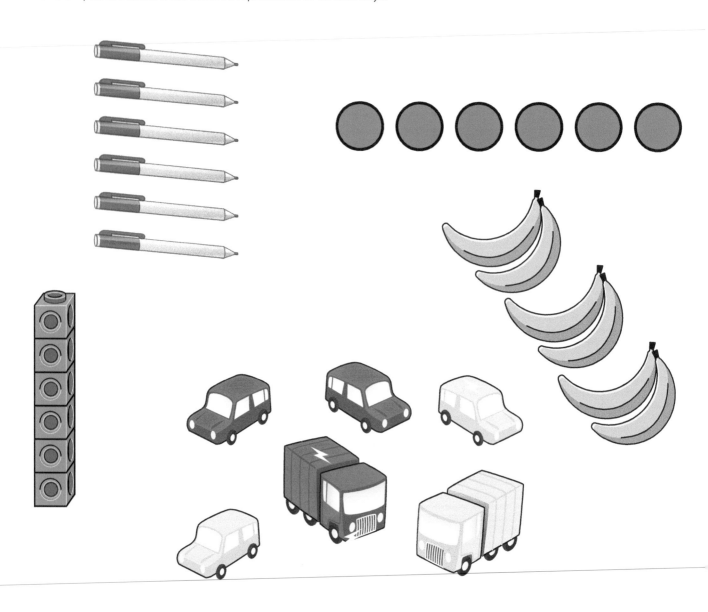

Practical aspects of *Power Maths*

One of the key underlying elements of *Power Maths* is its practical approach, allowing you to make maths real and relevant to your children, no matter their age.

Manipulatives are essential resources for both key stages and *Power Maths* encourages teachers to use these at every opportunity, and to continue the Concrete-Pictorial-Abstract approach right through to Year 6.

The Textbooks and Teacher Guides include lots of opportunities for teaching in a practical way to show children what maths means in real life.

Discover and Share

The **Discover** and **Share** sections of the Textbook give you scope to turn a real-life scenario into a practical and hands-on section of the lesson. Use these sections as inspiration to get active in the classroom. Where appropriate, use the **Discover** contexts as a springboard for your own examples that have particular resonance for your children – and allow them to get their hands dirty trying out the mathematics for themselves.

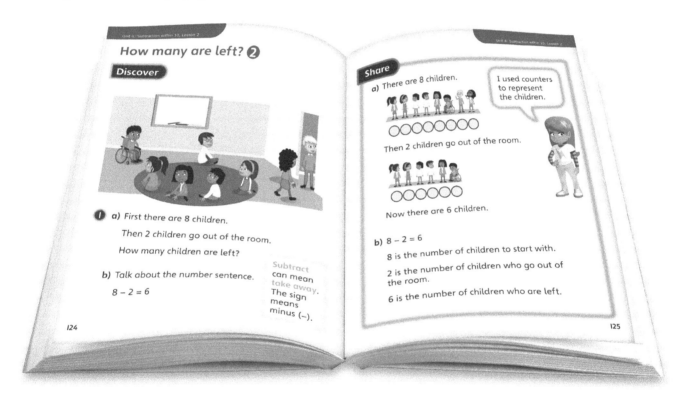

Unit videos

Every term has one unit video which incorporates real-life classroom sequences.

These videos show you how the reasoning behind mathematics can be carried out in a practical manner by showing real children using various concrete and pictorial methods to come to the solution. You can see how using these practical models, such as part-whole and bar models, helps them to find and articulate their answer.

Mastery tips

Mastery Experts give anecdotal advice on where they have used hands-on and real-life elements to inspire their children.

Mastery Expert tip! 'When I taught this unit, I used the characters within the Pupil Books to encourage children to talk about the different methods they were using. This worked well and helped us explore different ways of working. It also encouraged children to talk openly about mistakes.'

Don't forget to watch the Unit 3 video!

Concrete-Pictorial-Abstract (C-P-A) approach

Each **Share** section uses various methods to explain an answer, helping children to access abstract concepts by using concrete tools, such as counters. Remember, this isn't a linear process, so even children who appear confident using the more abstract method can deepen their knowledge by exploring the concrete representations. Encourage children to use all three methods to really solidify their understanding of a concept.

Pictorial representation – drawing the problem in a logical way that helps children visualise the maths

Concrete representation – using manipulatives to represent the problem. Encourage children to physically use resources to explore the maths.

Abstract representation – using words and calculations to represent the problem.

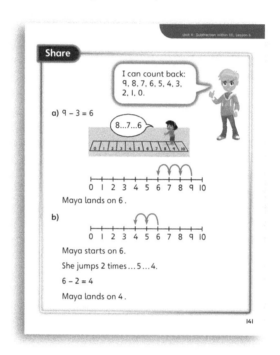

Practical tips

Every lesson suggests how to draw out the practical side of the **Discover** context.

You'll find these in the **Discover** section of the Teacher Guide for each lesson.

> **PRACTICAL TIPS** Ask children to use counters to represent the children that are standing up and sitting down.

Resources

Every lesson lists the practical resources you will need or might want to use. There is also a summary of all of the resources used throughout the term on page 40 to help you be prepared.

> **RESOURCES**
>
> **Mandatory:** part-whole model, counters/cubes
> **Optional:** ten frames, toys

Using *Power Maths* flexibly in Key Stage I

Power Maths lessons have a coherent, regular structure that supports you in building up children's understanding in a series of small steps. This is something most classes will need to build up to, rather than running in from a standing start at the beginning of Year 1.

Start by using the Practice Books in small groups

In most Year 1 classes, it won't be realistic for the whole class to complete the Practice Book pages independently at the start of the year, but they will learn to do this gradually. For the Textbooks, children will need to get used to direct teaching and recording answers in their own books. And, of course, this will set them up well for the rest of Primary school.

Small teacher-led groups are likely to be the best approach for independent practice. This format allows you to talk children through the question, discuss their ideas using manipulatives (often there will be manipulatives on the page as a hint), and guide them in representing their answer. (For instance, they can tell you the answer is 5, but they may need help writing 5 or knowing that they should colour in 5 apples.)

Go through the questions one-by-one with the group. You can mark their work/give feedback there and then. As children get used to the materials, the next stage could be for the small group to work through the questions at their own pace. The style of questions in *Power Maths* is quite regular, so children will get better at knowing what they need to do.

To facilitate small group work, you are likely to need some other activities as a carousel. A good way to do this is by turning a question from the Textbook into a game (usually **Think together** question 3 will work well) and teaching this to children before you break into groups. For instance, look at the example below (pages 78–79 in Textbook 1A). You could teach children a game with a part-whole model where one child puts in the whole using counters and the other children have to put in the parts. Or they could try this with beanbags and hoops. Base the practice on the key learning from the lesson.

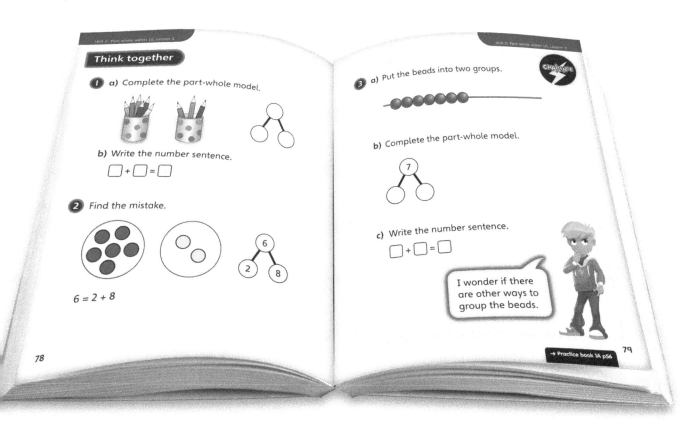

Are there any other ways to use the resources flexibly?

Don't be afraid to bring the **Discover** activity to life! Perhaps you could turn it into a game, or a role play. For instance, if the context is a teddy bear's picnic, you could share out fruit between teddies in the class. Or could you find a toy rocket to launch for the lesson below? (Textbook 1A page 32).

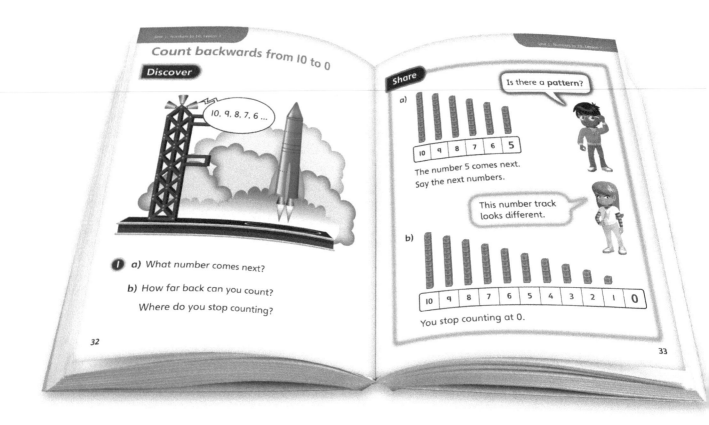

For some lessons you could consider a slightly different approach where you move backwards and forwards between the Textbook and Practice Book. If **Think together** question 1 links well with Practice Book question 1, you could do the **Think together** question together and then let children complete the Practice Book question, then the same for question 2, etc. This works better for some lessons than others, but it is one way of making practice more independent in short bursts, as a way of building up independence.

Don't forget, there isn't a *Power Maths* lesson for every lesson in the year. You can take more time where you need to, so that children's understanding is secure. In Key Stage 1, it will be all the more important to take your time, because children need to get used to the format as well as master the key learning. If using some of the ideas above means that a *Power Maths* lesson actually takes two lessons, e.g. for the first part of the year, then that's fine!

There are some further ideas for using the materials flexibly in the next section.

Working with children below age-related expectation

This section offers advice on using *Power Maths* with children who are significantly behind age-related expectation. Teacher judgement will be crucial in terms of where and why children are struggling, and in choosing the right approach. The suggestions can of course be adapted for children with special educational needs, depending on the specific details of those needs.

General approaches to support children who are struggling

Keeping the pace manageable

Remember, you have more teaching days than *Power Maths* lessons so you can cover a lesson over more than one day, and revisit key learning, to ensure all children are ready to move on. You can use the + and – buttons to adjust the time for each unit in the online planning. The NCETM's Ready-to-Progress criteria can be used to help determine what should be highest priority.

Same-day intervention

You could go over the Textbook pages or revisit the previous year's work if necessary. Remember that same-day intervention can be within the lesson, as well as afterwards (see page 29). As children start their independent practice, you can work with those who found the first part of the lesson difficult, checking understanding using manipulatives.

Fluency sessions

Fit in as much practice as you can for number bonds and times-tables, etc., at other times of the day. If you can, plan a short 'maths meeting' for this in the afternoon. You might choose to use a Power Up you haven't used already.

Pre-teaching

Find a 5- to 10-minute slot before the lesson to work with the children you feel would benefit. The afternoon before the lesson can work well, because it gives children time to think in between. Recap previous work on the topic (addressing any gaps you're aware of) and do some fluency practice, targeting number facts etc. that will help children access the learning.

Focusing on the key concepts

If children are a long way behind, it can be helpful to take a step back and think about the key concepts for children to engage with, not just the fine detail of the objective for that year group (e.g. addition with a specific number of columns). Bearing that in mind, how could children advance their understanding of the topic?

Providing extra support within the lesson

Support in the Teacher Guide

First of all, use the Strengthen support in the Teacher Guide for guided and independent work in each lesson, and share this with Teaching Assistants, where relevant. As you read through the lesson content and corresponding Teacher Guide pages before the lesson, ask yourself what key idea or nugget of understanding is at the heart of the lesson. If children are struggling, this should help you decide what's essential for all children before they move on.

Annotating pages

You can annotate questions to provide extra scaffolding or hints if you need to, but aim to build up children's ability to access questions independently wherever you can. Children tend to get used to the style of the *Power Maths* questions over time.

Quick recap as lesson starter

The Quick recap for each lesson in the Teacher Guide is an alternative starter activity to the Power Up. You might choose to use this with some or all children if you feel they will need support accessing the main lesson.

Consolidation questions

If you think some children would benefit from additional questions at the same level before moving on, write one or two similar questions on the board. (This shouldn't be at the expense of reasoning and problem-solving opportunities: take longer over the lesson if you need to.)

Hard copy Textbooks

The Textbooks help children focus in more easily on the mathematical representations, read the text more comfortably, and revisit work from a previous lesson that you are building on, as well as giving children ownership of their learning journey. In main lessons, it can work well to use the e-Textbook for **Discover** and give out the books when discussing the methods in the **Share** section.

Reading support

It's important that all children are exposed to problem solving and reasoning questions, which often involve reading. For whole-class work you can read questions together. For independent practice you could consider annotating pages to help children see what the question is asking, and stem sentences to help structure their answer. A general focus on specific mathematical language and vocabulary will help children access the questions. You could consider pairing weaker readers with stronger readers, or read questions as a group if those who need support are on the same table.

Providing extra depth and challenge with *Power Maths*

Just as prescribed in the National Curriculum, the goal of *Power Maths* is never to accelerate through a topic but rather to gain a clear, deep and broad understanding. Here are some suggestions to help ensure all children are appropriately challenged as you work with the resources.

Overall approaches

First of all, remember that the materials are designed to help you keep the class together, allowing all children to master a concept while those who grasp it quickly have time to explore it in more depth. Use the Deepen support in the Teacher Guide (see below) to challenge children who work through the questions quickly. Here are some questions and ideas to encourage breadth and depth during specific parts of the lesson, or at any time (where no part of the lesson sequence is specified):

- **Discover**: 'Can you demonstrate your solution another way?'
- **Share**: Make sure every child is encouraged to give answers and engage with the discussion, not just the most confident.
- **Think together**: 'Can you model your answers using concrete materials? Can you explain your solution to a partner?'
- Practice: Allow all children to work through the full set of questions, so that they benefit from the logical sequence.
- **Reflect**: 'Is there another way of working out the answer? And another way?'
  'Have you found all the solutions?'
  'Is that always true?'
  'What's different between this question and that question? And what's the same?'

Note that the **Challenge** questions are designed so that all children can access and attempt them, if they have worked through the steps leading up to them. There may be some children in a given lesson who don't manage to do the **Challenge**, but it is not supposed to be a distinct task for a subset of the class. When you look through the lesson materials before teaching, think about what each question is specifically asking, and compare this with the key learning point for the lesson. This will help you decide which questions you feel it's essential for all children to answer, before moving on. You can at least aim for all children to try the **Challenge**!

Deepen activities and support

The Teacher Guide provides valuable support for each stage of the lesson. This includes Deepen tips for the guided and independent practice sections, which will help you provide extra stretch and challenge within your lesson, without having to organise additional tasks. If you have a Teaching Assistant, they can also make use of this advice. There are also suggestions for the lesson as a whole in the 'Going Deeper' section on the first page of the Teacher Guide section for that lesson. Every class is different, so you can always go a bit further in the direction indicated, if appropriate, and build on the suggestions given.

There is a Deepen activity for each unit. These are designed to follow on from the End of unit check, stretching children who have a firm understanding of the key learning from the unit. Children can work on them independently, which makes it easier for the teacher to facilitate the Strengthen activity for children who need extra support. Deepen activities could also be introduced earlier in the unit if the necessary work has been covered. The Deepen activities are on *ActiveLearn* on the Planning page for each unit, and also on the Resources page).

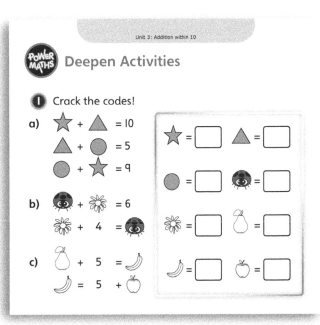

37

Using the questions flexibly to provide extra challenge

Sometimes you may want to write an extra question on the board or provide this on paper. You can usually do this by tweaking the lesson materials. The questions are designed to form a carefully structured sequence that builds understanding step by step, but, with careful thought about the purpose of each question, you can use the materials flexibly where you need to. Sometimes you might feel that children would benefit from another similar question for consolidation before moving on to the next one, or you might feel that they would benefit from a harder example in the same style. It should be quick and easy to generate 'more of the same' type questions where this is the case.

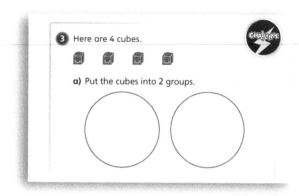

When you see a question like this one (from Unit 2, Lesson 1), it's easy to make harder examples to do afterwards if you need them. What if there were 9 cubes? Can children write the parts and wholes and find lots of different ways?

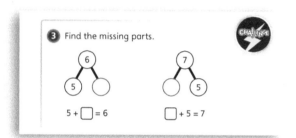

For this example (from Unit 3, Lesson 4), you could ask children to make up their own question(s) for a partner to solve. (In fact, for any of these examples you could ask early finishers to create their own question for a partner.)

Here's an example (from Unit 8, Lesson 2) where you could use the original context to provide extra challenge at the end of the lesson. For example, you could ask how far the frogs have to go to reach the next frog, or to be the winner.

Besides creating additional questions, you should be able to find a question in the lesson that you can adapt into a game or open-ended investigation, if this helps to keep everyone engaged. It could simply be that, instead of answering 5 + 6 etc. on the page, they could build a robot with 5 cubes and 6 cylinders.

> **Discover**
>
> ❶ a) Count to 50 as a class.
>
> b) What numbers are the frogs on?
>
> 112

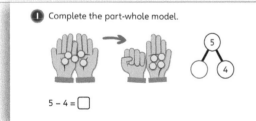

With a question like this one (from Unit 4, Lesson 4), children could play the game in pairs, taking an agreed number of counters and then showing what's in one hand. They could write a subtraction sentence each time, varying the whole and the parts.

See the bullets on the previous page for some general ideas that will help with 'opening out' questions in the books, e.g. 'Can you find all the solutions?' type questions.

Other suggestions

Another way of stretching children is through mixed ability pairs, or via other opportunities for children to explain their understanding in their own way. This is a good way of encouraging children to go deeper into the learning, rather than, for instance, tackling questions that are computationally more challenging but conceptually equivalent in level.

Using *Power Maths* with mixed age classes

Overall approaches

There are many variables between schools that would make it inadvisable to recommend a one-size-fits-all approach to mixed age teaching with *Power Maths*. These include how year groups are merged, availability of Teaching Assistants, experience and preference of teaching staff, range in pupil attainment across years, classroom space and layout, level of flexibility around timetables, and overall organisational structure (whether the school is part of a trust).

Some schools will find it best to timetable separate maths lessons for the different year groups. Others will aim to teach the class together as much as possible using the mixed age planning support on *ActiveLearn* (see the lesson exemplars for ways of organising lessons with strong/medium/weak correlation between year groups). There will also be ways of adapting these general approaches. For example, offset lessons where Year A start their lesson with the teacher, while Year B work independently on the practice from the previous lesson, and then start the next lesson with the teacher while Year A work independently; or teachers may choose to base their provision around the lesson from one year group and tweak the content up/down for the other group.

Key strategies for mixed age teaching

The mixed age teaching webinar on *ActiveLearn* provides advice on all aspects of mixed age teaching, including more detail on the ideas below.

Developing independence over time

Investing time in building up children's independence will pay off in the medium term.

Clear rationale

If someone asked, 'Why did you teach both Unit 3 and 4 in the same lesson/separate lessons?', what would your answer be?

Designing a lesson

1. Identify the core learning for each group
2. Identify any number skills necessary to access the core
3. Consider the flow of concepts and how one core leads to the other

Challenging all children

The questions are designed to build understanding step by step, but with careful thought about the purpose of each question you can tweak them to increase the challenge.

Multiple years combined

With more than two years together, teachers will inevitably need to use the resources flexibly if delivering a single lesson.

Enjoy the positives!

Comparison deepens understanding and there will be lots of opportunities for children, as well as misconceptions to explore. There is also in-built pre-teaching and the chance to build up a concept from its foundations. For teachers there is double the material to draw on! Mixed age teachers require a strong understanding of the progression of ideas across year groups, which is highly valuable for all teachers. Also, it is necessary to engage deeply with the lesson to see how to use the materials flexibly – this is recommended for all teachers and will help you bring your lesson to life!

List of practical resources

Year IA Mandatory resources

Resource	Lesson
2D shapes (squares and circles)	**Unit 1** Lesson 9
Bead string	**Unit 2** Lesson 6
Blocks	**Unit 2** Lesson 4
Colouring pencils	**Unit 1** Lessons 3, 4
Concrete 2D representations of squares, rectangles, triangles, circles	**Unit 5** Lessons 3, 5
Concrete 3D representations of cubes, cones, cuboids, cylinders, spheres, pyramids, cones, hemispheres	**Unit 5** Lessons 1, 2, 4, 5
Counters	**Unit 1** Lessons 1, 2, 3, 4, 5, 6, 7, 8, 9, 10, 11 **Unit 2** Lessons 1, 2, 3, 4, 5, 7 **Unit 3** Lessons 1, 4 **Unit 4** Lessons 1, 2, 4, 5, 6, 7
Counters (double-sided)	**Unit 4** Lesson 3
Cubes	**Unit 1** Lesson 3 **Unit 2** Lessons 1, 3, 4, 5, 7 **Unit 3** Lessons 1, 2, 4 **Unit 4** Lessons 1, 3, 5
Dice (six-sided)	**Unit 4** Lesson 6 **Unit 1** Lessons 7, 13
Digit cards	**Unit 1** Lesson 12 **Unit 3** Lesson 2 **Unit 4** Lessons 1, 7
Five frames	**Unit 2** Lesson 6
Jar	**Unit 3** Lesson 2
Marbles	**Unit 3** Lesson 2
Multilink cubes	**Unit 1** Lessons 1, 2, 4 5, 6, 7, 8, 9, 10, 11, 12, 13 **Unit 4** Lessons 2, 4, 6, 7
Number cards	**Unit 1** Lessons 2, 6, 7
Number lines	**Unit 1** Lesson 14 **Unit 3** Lessons 2, 3 **Unit 4** Lesson 6, 7
Number tracks	**Unit 1** Lessons 2, 6, 7, 8 **Unit 4** Lesson 6
Real-life objects (from the classroom)	**Unit 1** Lesson 1 **Unit 2** Lesson 1
Part-whole models	**Unit 2** Lesson 7 **Unit 3** Lessons 1, 3, 4 **Unit 4** Lessons 3, 4, 5, 7
Ten frames	**Unit 1** Lessons 3, 4, 5, 6, 7, 8, 9, 13 **Unit 2** Lessons 6, 7 **Unit 3** Lessons 1, 2 **Unit 4** Lesson 2
Whiteboard	**Unit 1** Lesson 2

Year IA Optional resources

Resource	Lesson
<, > and = signs (large printed versions) for children to hold and manipulate	**Unit 1** Lessons 11, 12
+, – and = signs (printed)	**Unit 4** Lesson 5
2D shapes (squares and circles)	**Unit 1** Lesson 10
3D shape name labels	**Unit 5** Lessons 2, 4
Bag (opaque)	**Unit 5** Lesson 1
Balloons (or pictures of)	**Unit 4** Lesson 1
Bead strings	**Unit 1** Lessons 5, 12 **Unit 2** Lessons 5, 7 **Unit 3** Lesson 2
Cardboard tube (or other item to be used as a rocket)	**Unit 1** Lesson 7
Chalk	**Unit 1** Lesson 14 **Unit 2** Lesson 3
Concrete resources to model addition calculations (e.g., cubes, counters, teddies, cars)	**Unit 3** Lesson 1
Countable real-life objects	**Unit 1** Lessons 2, 3, 4, 6, 7, 8, 9, 10, 11, 12, 14 **Unit 2** Lesson 2
Counters	**Unit 1** Lesson 12 **Unit 2** Lesson 6 **Unit 4** Lesson 7
Craft paper	**Unit 4** Lesson 7
Cubes	**Unit 2** Lesson 6 **Unit 4** Lesson 7
Cups or glasses of juice or water	**Unit 2** Lesson 4
Dice (six-sided)	**Unit 1** Lessons 5, 14
Dice (ten-sided)	**Unit 1** Lesson 12
Dry-wipe markers	**Unit 5** Lesson 4
Egg boxes	**Unit 4** Lesson 4
Everyday items relating to the 3D shapes (e.g., golf ball, cereal box, empty sweet tube, dice)	**Unit 5** Lessons 1, 2
Flashcards (of numbers 1 to 10)	**Unit 1** Lesson 14
Fruit (real or play)	**Unit 4** Lesson 4
Groupable 2D and 3D shapes	**Unit 1** Lesson 1
Hoops	**Unit 2** Lessons 2, 3, 4
Hoops (small) and a post or bean bags and a large hoop	**Unit 4** Lesson 5
Hopscotch grid	**Unit 4** Lesson 6
Language and comparison signs (classroom display)	**Unit 1** Lesson 12
Marker pens	**Unit 4** Lesson 7
Modelling material (to model 3D shapes)	**Unit 5** Lesson 1
Multilink cubes	**Unit 1** Lesson 14 **Unit 3** Lesson 2
Number cards	**Unit 1** Lesson 5 **Unit 4** Lesson 5
Number cards (with examples of the concrete materials used in the lesson)	**Unit 1** Lesson 13
Magnetic numbers	**Unit 4** Lesson 5
Number fan	**Unit 1** Lesson 2

Year IA Optional resources – *continued*

Resource	Lesson
Number line (giant)	**Unit 4** Lesson 6
Number sentence scaffolds (printed)	**Unit 4** Lesson 5
Number track (large) on display in the classroom	**Unit 1** Lesson 13
Number tracks	**Unit 1** Lessons 5, 9, 10, 11, 14 **Unit 3** Lesson 2
Paint and paper for shape-printing	**Unit 5** Lessons 4, 5
Pegs	**Unit 1** Lesson 14
Plastic food and flowers	**Unit 2** Lesson 4
Real-life objects (classroom, e.g., pencils and pencil pots)	**Unit 2** Lesson 3
Real-life objects (classroom/PE equipment)	**Unit 3** Lesson 3
Rekenrek	**Unit 1** Lesson 5 **Unit 2** Lessons 5, 6
Sorting hoops	**Unit 5** Lessons 1, 3, 4
String	**Unit 1** Lesson 14 **Unit 2** Lesson 3
Teddy bears	**Unit 2** Lessons 2, 4
Ten frames	**Unit 1** Lesson 2, 10, 12, 14 **Unit 2** Lesson 5 **Unit 3** Lesson 4 **Unit 4** Lesson 4
Toys	**Unit 2** Lesson 4 **Unit 3** Lesson 4
Toys (cars)	**Unit 4** Lesson 3

Getting started with *Power Maths*

As you prepare to put *Power Maths* into action, you might find the tips and advice below helpful.

STEP 1: Train up!

A practical, up-front full day professional development course will give you and your team a brilliant head-start as you begin your *Power Maths* journey. You will learn more about the ethos, how it works and why.

STEP 2: Check out the progression

Take a look at the yearly and termly overviews. Next take a look at the unit overview for the unit you are about to teach in your Teacher Guide, remembering that you can match your lessons and pacing to your class.

STEP 3: Explore the context

Take a little time to look at the context for this unit: what are the implications for the unit ahead? (Think about key language, common misunderstandings and intervention strategies, for example.) If you have the online subscription, don't forget to watch the corresponding unit video.

STEP 4: Prepare for your first lesson

Familiarise yourself with the objectives, essential questions to ask and the resources you will need. The Teacher Guide offers tips, ideas and guidance on individual lessons to help you anticipate children's misconceptions and challenge those who are ready to think more deeply.

STEP 5: Teach and reflect

Deliver your lesson – and enjoy!

Afterwards, reflect on how it went … Did you cover all five stages? Does the lesson need more time? How could you improve it?

Unit 1
Numbers to 10

Don't forget to watch the Unit 1 video!

Mastery Expert tip! 'When teaching this unit, I used the contexts given in the pictures to make the maths as practical as possible. The children were far more confident about explaining their ideas when we were role-playing the concepts or building them with resources.'

WHY THIS UNIT IS IMPORTANT

This unit focuses on children's ability to recognise, represent and manipulate numbers to 10. Children begin by practising and developing their ability to sort and group objects using different criteria, then move on to counting groups of objects up to 10. Children will learn to recognise and count different representations of numbers to 10 and use a ten frame to help structure counting and reasoning.

As children become more confident with counting they will be introduced to the appropriate vocabulary of counting: the word 'digit' and the written names of each number. They will move on to counting backwards and recognising 'one more' as a number increasing and 'one less' as a number decreasing.

Children will use all these skills to compare and order numbers to 10, using concrete and pictorial representations to support their reasoning. Finally, they will learn about ordinal numbers and be introduced to the number line as a representation of counting one more or one less.

WHERE THIS UNIT FITS

→ **Unit 1: Numbers to 10**
→ Unit 2: Part-whole within 10

In this unit, children begin by sorting and grouping objects up to 10. They then count to 10 and focus on 'one more' and 'one less' before learning how to use a number line to count forwards and backwards.

Before they start this unit, it is expected that children can:
• describe similarities and differences between objects
• sort objects into groups based on simple criteria.

ASSESSING MASTERY

Children who have mastered this unit will be able to confidently count forwards and backwards to and from 10. They will be able to recognise one more and one less than a number up to 10 and will be able to represent this using concrete, pictorial and abstract representations; they will use this understanding to correctly compare and order numbers.

COMMON MISCONCEPTIONS	STRENGTHENING UNDERSTANDING	GOING DEEPER
Children may find counting backwards trickier and mistakenly count forwards instead.	Role-play situations where counting down is necessary, such as a rocket launch or blowing out birthday candles, following the count on a number line or number track. Sing songs like 'Ten Green Bottles'.	Ask children to investigate how many times they need 'one more' or 'one less' to get from one number to another. Ask: *What representations could you use to show this?*
Children think that the number 5 is represented in just one way on a ten frame.	Discuss with children how to make 5: using double-sided counters, can they see it how might be made up as a 3 and 2? Look at different ways of representing 5, such as on a rekenrek or as towers of cubes.	It is important to show children that they can represent 5 in many different ways. They may use a five frame or ten frame to do this. Ask: *Can you represent 5? Can you do it in a different way?*

Unit I: Numbers to 10

Use these pages to introduce the unit focus to children. You can use the characters to explore different ways of working too.

STRUCTURES AND REPRESENTATIONS

Ten frame: This model will help children visualise 10. It will also help strengthen children's fluency with numbers up to 10, demonstrating how they can be arranged in different ways but still be worth the same amount.

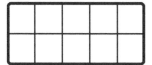

Number track: This model will help children organise their representations of numbers from 1 upwards. It can help children with comparing and ordering numbers.

Number line: This model helps children visualise the order of numbers. It can help them demonstrate concepts such as 'one more' and 'one less' in a more efficient way than using concrete resources.

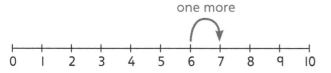

KEY LANGUAGE

There is some key language that children will need to know as a part of the learning in this unit:

→ sort, groups, pattern
→ digit, number
→ count on, count back, count up, one more, one more than, one less, one less than
→ matched, equal to, =
→ fewer, less than, <, least, fewest, smallest
→ more, greater than, >, most, greatest
→ number line, number track, ten frame

PUPIL TEXTBOOK 1A PAGE 6

PUPIL TEXTBOOK 1A PAGE 7

45

Sort objects

Learning focus

In this lesson, children will develop their understanding of grouping objects. They will be able to recognise and explain different ways of sorting objects and that a single group of objects can be sorted in multiple ways.

Before you teach

- Are all children able to confidently explain similarities and differences between two or more objects?
- How could you add challenge to the process of grouping objects? For example, could all objects be the same in some way so that children have to use more than one criterion?

NATIONAL CURRICULUM LINKS

Year 1 Number – number and place value
Identify and represent numbers using objects and pictorial representations including the number line, and use the language of: equal to, more than, less than (fewer), most, least.

ASSESSING MASTERY

Children can group objects based on their similarities and differences. Children can recognise that the same group of objects can be grouped in different ways by changing the grouping criteria.

COMMON MISCONCEPTIONS

Children may only want to group objects by one criterion with which they are most comfortable, such as colour. To elicit more criteria, ask:
- *What else is the same about the objects?*
- *What else is different about the objects?*

STRENGTHENING UNDERSTANDING

Ask children to carefully group real-life objects, based on different criteria that you provide. Ask: *Can you group all the cubes? Can you group all the circles?*

GOING DEEPER

Challenge children to find as many different ways of grouping the same objects as possible. Ask children to justify their criteria each time and explain why they have chosen those criteria. Children could draw their groupings and write labels for each grouping.

KEY LANGUAGE

In lesson: sort, groups

Other language to be used by the teacher: same, different

RESOURCES

Mandatory: counters, multilink cubes, a selection of groupable real-life objects such as toy cars, pens and pencils

Optional: other groupable 2D and 3D shapes

 In the eTextbook of this lesson, you will find interactive links to a selection of teaching tools.

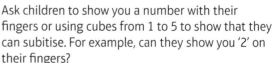

Quick recap 🔁

Ask children to show you a number with their fingers or using cubes from 1 to 5 to show that they can subitise. For example, can they show you '2' on their fingers?

Discover

Unit 1: Numbers to 10, Lesson 1

Sort objects

Discover

WAYS OF WORKING Pair work

ASK

- Question **1** a): *Does the colour of the multilink cubes and the counters matter? Why?*
- Question **1** b): *Why have you sorted the fruit in different ways? Are some ways more useful than others? Why or why not?*

IN FOCUS Question **1** a) introduces all of the lesson's key language, supporting children in understanding that different objects can be grouped together.

Question **1** b) allows children to begin designing their own criteria. This question allows paired and class discussion about how and why children have sorted the fruit differently.

PRACTICAL TIPS Use counters, cubes and real-life objects to recreate the scenario in the classroom.

ANSWERS

Question **1** a): There is a group of cubes and a group of counters. There is also a group of red objects and a group of yellow objects.

Question **1** b): The fruit can be sorted in two different ways. The first way is that it can be sorted into three groups: a group of two apples, a group of two oranges and a group of three bananas. The second way is that it could also be sorted into two groups: a group of four round fruit and a group of three non-round fruit.

1 a) Sort the ◯ and 🔲 into two groups.

b) Sort the fruit. What groups did you make?

8

PUPIL TEXTBOOK 1A PAGE 8

Share

WAYS OF WORKING Whole class teacher led

ASK

- *Do you think sorting the fruit into two groups works? Can you explain why?*
- *Can you sort the fruit in a different way?*
- *How have you made your groups clear?*

IN FOCUS Question **1** a) demonstrates how a single group of objects can be sorted in more than one way. Use this opportunity to approach the potential misconception that objects can only be sorted using one criterion. Discuss how and why the groups change depending on the way in which the objects are sorted.

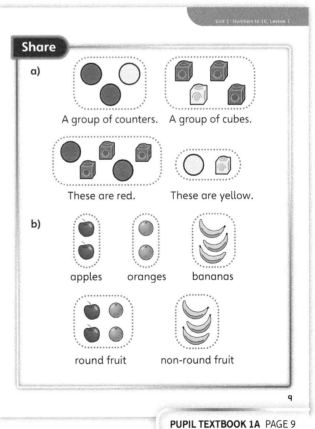

Think together

WAYS OF WORKING Whole class teacher led (I do, We do, You do)

ASK

- *What can you see?*
- *How are the objects the same or different?*
- *How will you know how to group the objects?*
- *How will you show the groups you have made?*

IN FOCUS Question ❶ and question ❷ allow children to begin sorting objects by themselves, asking them to group objects by circling the pictorial representations of the objects.

Question ❸ gives children the opportunity to group similar objects in different ways based on their own criteria. This question provides opportunities to discuss different children's criteria, looking for their reasoning and their ability to justify their choices. It is important that children recognise that the objects can be sorted in a number of different ways.

STRENGTHEN Give children a selection of real-life objects to sort. You could give children the criteria to use when sorting these objects or children could come up with their own criteria.

DEEPEN Refer to Ash's question about whether there are more ways to group the objects. This will prompt children to consider alternative criteria for their groupings. Ask children to sort the objects into three different groups.

ASSESSMENT CHECKPOINT Question ❸ assesses if children are able to confidently choose criteria by which to group objects and then group those objects successfully. It will also assess if children recognise how a set of objects can be grouped in more than one way.

ANSWERS

Question ❶: A group of pens and a group of counters.

Question ❷: A group of bananas and a group of cherries.

Question ❸: A group of cars and a group of trucks, or a group of vehicles with a white flash and a group of vehicles without a white flash, or a group of red vehicles and a group of yellow vehicles.

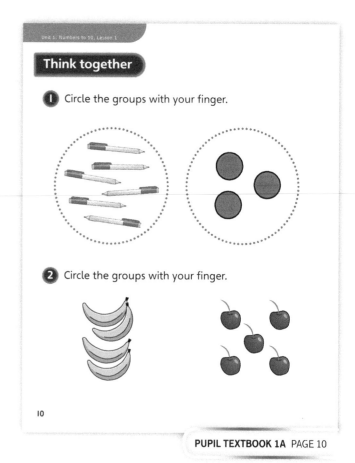

PUPIL TEXTBOOK 1A PAGE 10

PUPIL TEXTBOOK 1A PAGE 11

Practice

WAYS OF WORKING Independent thinking

IN FOCUS Questions **1**, **2** and **3** provide further support and develop children's ability to independently find, sort and group objects based on their own criteria.

STRENGTHEN Ask children to sort a selection of real-life objects. Depending on children's needs, you could give them criteria to use when sorting the objects or ask them to come up with their own criteria.

DEEPEN Question **5** supports children in finding different ways of sorting a group of objects. Listen for children's explanations of how they have sorted and grouped the objects. Do they recognise that there were other ways of sorting? Can they tell you how many other ways they can see? Can they sort the objects using three criteria?

THINK DIFFERENTLY Question **4** requires children to think differently because they have to sort the objects that they think belong in groups in order to identify the object that does not belong.

ASSESSMENT CHECKPOINT Questions **1**, **2** and **3** should help you to assess if children are able to independently sort and group objects according to their own criteria.

ANSWERS Answers for the **Practice** part of the lesson can be found in the *Power Maths* online subscription.

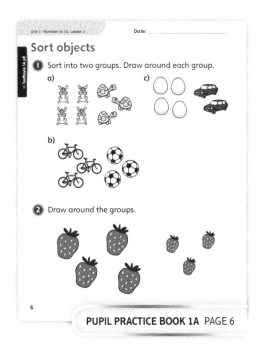

PUPIL PRACTICE BOOK 1A PAGE 6

PUPIL PRACTICE BOOK 1A PAGE 7

Reflect

WAYS OF WORKING Independent thinking

IN FOCUS This **Reflect** activity gives children the opportunity to put into practice all the skills that they have developed in this lesson. Encourage children to discuss their reasoning and justify their choices with their partner.

ASSESSMENT CHECKPOINT This **Reflect** activity should help you assess if children can choose objects that are in some way similar to one another. Assess if children can group their objects based on their chosen criteria.

ANSWERS Answers for the **Reflect** part of the lesson can be found in the *Power Maths* online subscription.

After the lesson ⏸

- Are children more confident in recognising that objects can be sorted and grouped in many different ways?
- How has the lesson enabled children to explain their criteria for sorting the objects and justify their ideas?
- Could children's learning in this lesson be supported and developed through another area of the curriculum?

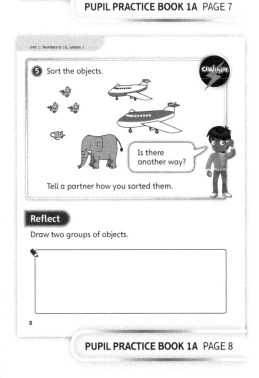

PUPIL PRACTICE BOOK 1A PAGE 8

Count objects to 10

Learning focus

In this lesson, children will be able to relate the amount of objects to the correct number in digits and the correct number in words.

Before you teach

- Do children confidently recognise and understand each number up to 10?

NATIONAL CURRICULUM LINKS

Year 1 Number – number and place value

Count to and across 100, forwards and backwards, beginning with 0 or 1, or from any given number.

Identify and represent numbers using objects and pictorial representations including the number line, and use the language of: equal to, more than, less than (fewer), most, least.

ASSESSING MASTERY

Children can count objects accurately and link the amount to the correct numeral and word. Children can recognise that countable objects may be represented in many different ways, not just as regularly ordered rows of single objects.

COMMON MISCONCEPTIONS

Children may count too many or too few. Counting the same object more than once is common. Ask:
- *What could you use to help you count the objects?*

Children may relate the size of the objects to the amount of the objects. For example, when comparing two elephants to two mice, they may suggest there are more elephants because they are larger in size. Ask:
- *Can you replace the objects with counters?*
- *Has your answer changed?*

STRENGTHENING UNDERSTANDING

Give children number cards, each of which should show a numeral from 1 to 10, the corresponding number in words and the corresponding number in pictures. These will support understanding throughout this lesson.

Children could role-play going to the shops with a shopping list and buying the correct number of items. This reinforces understanding of the link between an abstract number and a concrete amount.

GOING DEEPER

Ask children to show how many different ways they can represent a given number. Ask: *How many ways can you arrange the milk cartons to make 7?*

KEY LANGUAGE

In lesson: one, two, three, four, five, six, seven, eight, nine, ten

Other language to be used by the teacher: different, same

STRUCTURES AND REPRESENTATIONS

Number tracks

RESOURCES

Mandatory: counters, multilink cubes, number cards, number tracks, whiteboard

Optional: a selection of countable real-life objects from the classroom, ten frames, number fan

 In the eTextbook of this lesson, you will find interactive links to a selection of teaching tools.

Quick recap

Say a number from 0 to 10 out loud. Ask children to write that number digit on a whiteboard or show you with a number fan. It's important that children recognise the numeral for the word.

Discover

Unit 1: Numbers to 10, Lesson 2

Count objects to 10

WAYS OF WORKING Pair work

ASK

- Question ❶ a): *Before you count the objects, can you predict how many are in each group?*
- Question ❶ a): *How are you going to count the objects so you don't miss any out?*
- Question ❶ b): *How many boxes of cereal are there?*

IN FOCUS Questions ❶ a) and ❶ b) ask children to count the number of objects and record the number in two different ways. They start to look at efficient counting strategies so they don't miss any out, for example, they might count all the top row and then the bottom row. They then look at writing numbers as numerals and in words.

PRACTICAL TIPS Set up a food shop in the classroom with cans and cereal boxes displayed for counting. Or, use ten of the same real-life objects to represent the cans to recreate the scenario.

ANSWERS

Question ❶ a): There are 10 cans.

Question ❶ b): The number is 10. The word is ten.

Discover

❶ a) Count the 🥫.

b) Write your answer in two ways.

12

PUPIL TEXTBOOK 1A PAGE 12

Share

WAYS OF WORKING Whole class teacher led

ASK

- Question ❶ a): *How did you show 10? Why did you show it in that way?*
- Question ❶ a): *Did anyone else show it in a different way?*
- Question ❶ b): *Can you explain how the two representations are the same and how they are different?*
- Question ❶ b): *Can you think of another way to show 10?*
- Question ❶ b): *How could you show the number of cereal boxes?*

IN FOCUS Questions ❶ a) and ❶ b) introduce children to the written word representation of the numbers 1 to 10. They also provide children with different visual arrangements of the numbers 1 to 10, beginning to show the comparative size of each number.

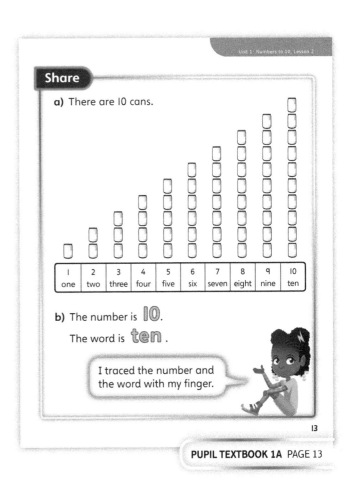

PUPIL TEXTBOOK 1A PAGE 13

Think together

Think together

WAYS OF WORKING Whole class teacher led (I do, We do, You do)

ASK

- *What do you need to count?*
- *How could you check your answer?*
- *How can you show your answer?*
- *Which arrangement is easiest to count? Explain your ideas.*
- *Could you show the numbers in a different way?*
- *How many different ways could you show the numbers?*

IN FOCUS Children count objects, starting with objects set out in lines as if they were on a ten frame in question ❶, to objects set out more randomly and in different orientations in questions ❷ and ❸. Question ❸ a) starts to look at how children can count efficiently making sure they don't miss any out and question ❸ b) provides an example where children have to group objects before counting.

STRENGTHEN Children put a counter over each of the objects and then count the counters. This ensures they don't miss any out. Alternatively, provide them with real-life objects or concrete materials to represent them. Ask children to arrange the objects or materials as they are shown in the pictures. Then arrange them either on a ten frame or a number track to help them count.

DEEPEN Ask children how many different ways they can count the objects in Question ❸ a) ask: *Is one way easier than the other?* Ask children how they might help someone who is struggling with Question ❸ b). What advice would they give?

ASSESSMENT CHECKPOINT Question ❸ should help you to assess if children are counting accurately and confidently, particularly when objects have been arranged differently or are of different sizes.

ANSWERS

Question ❶: There are 7 crisp packets.

Question ❷: There are 3 bananas and 5 apples.

Question ❸ a): There are 9 cereal boxes.

Question ❸ b): There are 6 oranges and 5 apple juices.

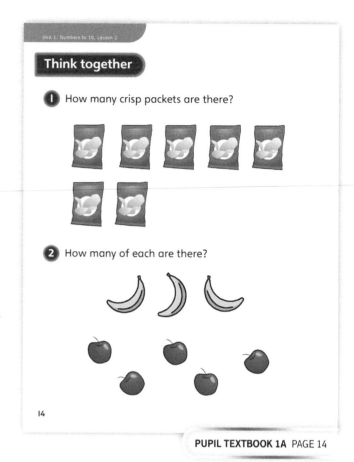

PUPIL TEXTBOOK 1A PAGE 14

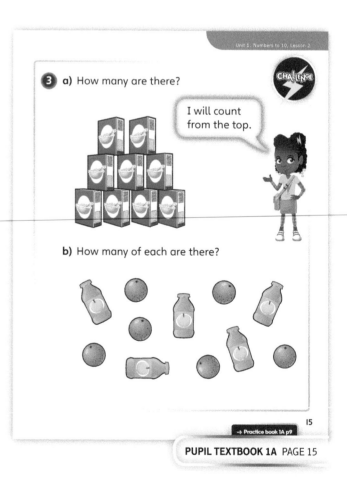

PUPIL TEXTBOOK 1A PAGE 15

Practice

WAYS OF WORKING Independent thinking

IN FOCUS Children learn to count objects to 10. In question ❶, they begin counting the same object set out in lines before moving on to counting different objects within the same group set out randomly in question ❷. Question ❸ provides an example where children have objects that are mixed up. They may cross out the objects as they count them or use counters to cover the objects and then count. Look out for children who just count all the objects.

STRENGTHEN Provide children with concrete objects to represent each question. Where objects are set out randomly, encourage children to cross out as they count or use counters to cover the objects and then count these.

DEEPEN Encourage children to subitise (recognise a small number of objects instantly) rather than count one by one. Question ❻ requires children to take note of their starting point.

THINK DIFFERENTLY Question ❺ asks children to count the number of objects in two groups and recognise that the objects are in two groups.

ASSESSMENT CHECKPOINT Questions ❶ to ❺ should help you assess children's ability to count objects reliably, regardless of size, orientation or arrangement.

ANSWERS Answers for the **Practice** part of the lesson can be found in the *Power Maths* online subscription.

Reflect

WAYS OF WORKING Independent thinking

IN FOCUS This **Reflect** activity allows children to reflect on the numbers they have learnt about and reinforce their ability to order numbers and write numbers in digits. Children should be able to confidently pick a number, record the related digit and offer more than one representation of it.

ASSESSMENT CHECKPOINT Assess if children use the correct digit to represent each number. Check that they do not miss any numbers out in the count and that the numbers are all in the correct order. Count back from 10, using the digits to check.

ANSWERS Answers for the **Reflect** part of the lesson can be found in the *Power Maths* online subscription.

After the lesson

- Can children reliably count objects, regardless of the objects' size, orientation and arrangement?
- Can all children confidently recognise both the digit and written name of every number from 1 to 10?

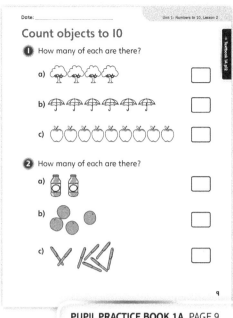

Date: _____

Count objects to 10

❶ How many of each are there?

a)

b)

c)

❷ How many of each are there?

a)

b)

c)

PUPIL PRACTICE BOOK 1A PAGE 9

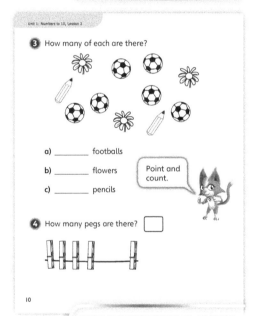

Unit 1: Numbers to 10, Lesson 2

❸ How many of each are there?

a) _____ footballs

b) _____ flowers

c) _____ pencils

Point and count.

❹ How many pegs are there? ☐

10

PUPIL PRACTICE BOOK 1A PAGE 10

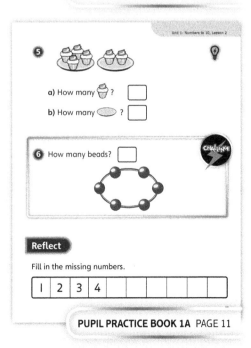

Unit 1: Numbers to 10, Lesson 2

❺

a) How many 🧁 ? ☐

b) How many 🍪 ? ☐

❻ How many beads? ☐

CHALLENGE

Reflect

Fill in the missing numbers.

1	2	3	4						

PUPIL PRACTICE BOOK 1A PAGE 11

Represent numbers to 10

Learning focus

In this lesson, children represent numbers using objects such as cubes or counters. The focus is now on more abstract objects rather than pictures or objects of real-life things.

Before you teach

- How will you ensure you explicitly show children that they can use counters or cubes to represent numbers or real-life objects?
- How will you support children who are struggling to see this link?
- How will you reinforce the learning from Lesson 2 on counting objects to 10?

NATIONAL CURRICULUM LINKS

Year 1 Number – number and place value

Count to and across 100, forwards and backwards, beginning with 0 or 1, or from any given number.

Identify and represent numbers using objects and pictorial representations including the number line, and use the language of: equal to, more than, less than (fewer), most, least.

ASSESSING MASTERY

Children can correctly represent numbers using objects. They understand that one cube can represent the number 1, two cubes can represent the number 2, etc. Children will represent these numbers on a ten frame.

COMMON MISCONCEPTIONS

Children may get confused that they can use a counter or a cube to represent a real-life object like a cake or an apple. Make the link explicit by replacing each real-life object with a cube or counter. Ask:
- *Replace each real-life object with a cube or a counter. What do you notice?*

STRENGTHENING UNDERSTANDING

Use real-life objects and cubes or counters side by side to show that they can represent the same number. Talk to children about why it might not be practical to always use the real-life objects. Give them a choice of what colour cube or counter they can use to represent the real-life objects to help them engage with the questions.

GOING DEEPER

Ask children if it matters what colour they use to represent real-life objects. Ask: *Do all the cubes or counters need to be the same colour? Does it make a difference?*

KEY LANGUAGE

In lesson: count, how many?

Other language to be used by the teacher: represent

STRUCTURES AND REPRESENTATIONS

Ten frames

RESOURCES

Mandatory: ten frames, cubes, counters, colouring pencils

Optional: other real-life objects that children can replace using cubes or counters

 In the eTextbook of this lesson, you will find interactive links to a selection of teaching tools.

Quick recap

Ask children to count a number of objects (up to 10) that are on the table or in the classroom. For example, count posters on the wall.

Discover

Represent numbers to 10

WAYS OF WORKING Pair work

ASK

- Question ① a): *How will you count the cubes? Will a ten frame help?*
- Question ① b): *Does the number change if we use counters instead of cubes?*

IN FOCUS Question ① b) introduces children to the fact that they can represent an object with a different object, such as counters, but the number they have remains the same. Children will then represent the number of cubes on the ten frame using counters. They may put the cubes on first and then replace with counters. They will notice that the numbers are the same.

PRACTICAL TIPS Count the cubes by placing them on a ten frame. Then one by one replace each cube with a counter.

ANSWERS

Question ① a): There are 7 cubes.

Question ① b): Children may fill the ten frame in different ways.

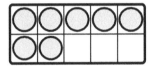

Discover

① **a)** Count the cubes.

    b) Show this number using counters on a ten frame.

16

PUPIL TEXTBOOK 1A PAGE 16

Share

WAYS OF WORKING Whole class teacher led

ASK

- Question ① a): *Why has Dexter lined the cubes up? Let's count out loud as a class to work out how many cubes there are. Point and count.*
- Question ① b): *Sparks has used a ten frame and counters. How does this help Sparks to count? How many counters are on the top row? Do we need to start counting at 1?*

IN FOCUS In question ① b), children are exposed to representing the same number using different objects, whilst also counting them in different ways. Children may have a preferred way of counting. Ask: *Is one way better than the other?*

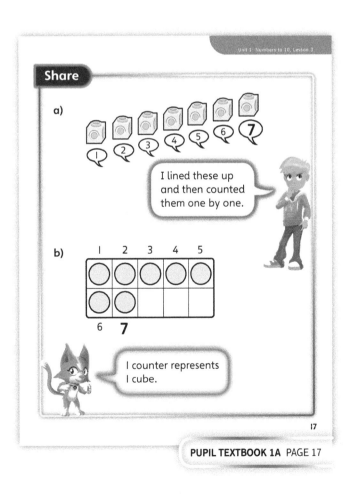

PUPIL TEXTBOOK 1A PAGE 17

Think together

Think together

WAYS OF WORKING Whole class teacher led (I do, We do, You do)

ASK

- Question ❶: *What could we use to represent the apples? How many apples can you see? Do you need to count? Does it matter whether you use cubes, counters or something else?*
- Question ❷: *Are the cubes easier to count in the tower or on the ten frame? Why?*

IN FOCUS In questions ❶ and ❷, children use counters or cubes on a ten frame to represent each image. Ask them whether the colour of their counters or cubes matter. Children may also just be able to see that there are 5 apples or 6 cubes when represented on a ten frame. This is subitising and something that should be encouraged. Ask: *Does it matter whether you use cubes or counters to represent the fruit?*

STRENGTHEN For question ❷, build the number using counters on a ten frame. Start by putting the cubes on the ten frame and replacing them with counters. Children should point and say the number out loud as they count. Encourage children to fill up the top row first and then the second row to aid counting.

DEEPEN Ask children to build numbers on a ten frame in the standard way (left to right and top row first) then ask them to build it in a different way. Ask: *Is the amount the same? How do you know? Which way is easier to count? Why?*

ASSESSMENT CHECKPOINT Questions ❶ to ❸ should help you assess whether children understand that they can represent numbers using counters or cubes and that colour and type are irrelevant. Question ❸ also helps children reinforce the formation of numbers on a ten frame, they may begin to start instantly recognising them (subitising).

ANSWERS

Question ❶: There are 5 apples.

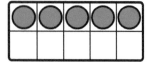

Question ❷: There are 6 cubes.

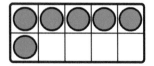

Question ❸: 1, 2, 3, 4, 5, 6.

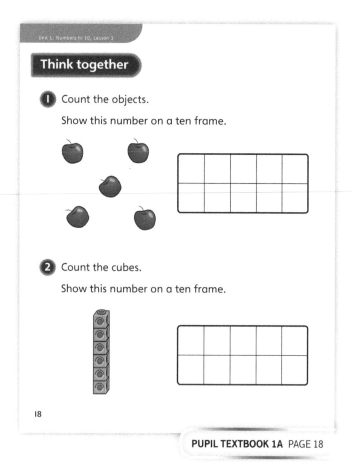

Unit 1: Numbers to 10, Lesson 3

Think together

❶ Count the objects.

Show this number on a ten frame.

❷ Count the cubes.

Show this number on a ten frame.

18

PUPIL TEXTBOOK 1A PAGE 18

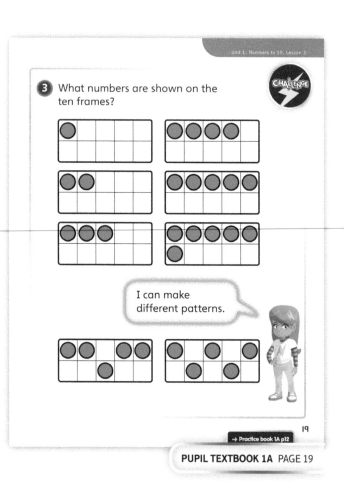

Unit 1: Numbers to 10, Lesson 3

❸ What numbers are shown on the ten frames?

CHALLENGE

I can make different patterns.

19

→ Practice book 1A p12

PUPIL TEXTBOOK 1A PAGE 19

Practice

WAYS OF WORKING Independent thinking

IN FOCUS Question ❶ focusses on children recognising numbers represented on ten frames. Question ❷ builds on question ❶ by asking children to count the objects, write the numeral, then select and shade in the correct number of counters. In question ❸, children draw their own counters to represent the numbers.

STRENGTHEN Use counters and ten frames so that children can represent each question physically.

DEEPEN Ask children to place their counters in a variety of ways. Ask: *Have you still got the same answer as your partner?*

ASSESSMENT CHECKPOINT Question ❸ in particular is useful to see if children understand how to represent numbers in depth.

ANSWERS Answers for the **Practice** part of the lesson can be found in the *Power Maths* online subscription.

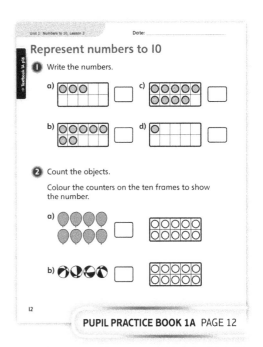

PUPIL PRACTICE BOOK 1A PAGE 12

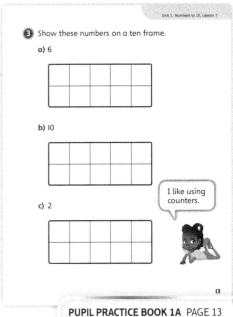

PUPIL PRACTICE BOOK 1A PAGE 13

Reflect

WAYS OF WORKING Independent thinking

IN FOCUS This **Reflect** activity allows children to reflect on the numbers they have learnt about and pick one they would like to examine further. To expose any misconceptions, you could give children a second number to focus on after they have worked on their own choice of number.

ASSESSMENT CHECKPOINT Assess if children are using the correct number to represent an amount. Have they used the ten frame to represent the number? What other ways are they able to represent it?

ANSWERS Answers for the **Reflect** part of the lesson can be found in the *Power Maths* online subscription.

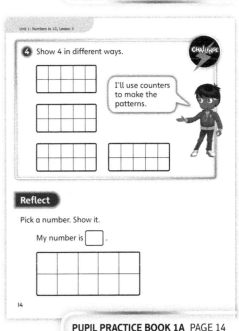

PUPIL PRACTICE BOOK 1A PAGE 14

After the lesson ⏸

- Are all children able to represent numbers to 10 in a concrete and an abstract manner?
- Are you confident that all children know what each number looks like as a concrete amount of objects?
- How will you use the representations introduced in this lesson to support children in the future?

Count objects from a larger group

Learning focus

In this lesson, children will count objects from a larger group and will need to be secure with the counting skill of one-to-one correspondence before attempting this lesson. This is one of the first times children will also need to think about the way they organise their resources.

Before you teach

- Are children confident counting all objects in a group?
- How can you help children organise their resources?

NATIONAL CURRICULUM LINKS

Year 1 Number – number and place value

Count to and across 100, forwards and backwards, beginning with 0 or 1, or from any given number.

Identify and represent numbers using objects and pictorial representations including the number line, and use the language of: equal to, more than, less than (fewer), most, least.

ASSESSING MASTERY

Children can count any number up to 10 from a larger group of objects and can organise their learning appropriately.

COMMON MISCONCEPTIONS

Children may start counting from 0 rather than 1 when using objects and therefore get an answer that is one more than what they should have. Reinforce the meaning of zero. Ask:
- *Would there be any objects if there were zero objects?*
- *When would you say the number zero? Would you say it before or after counting your first object?*

STRENGTHENING UNDERSTANDING

Encourage children to physically move objects one-by-one and count them as they move each object. They could place each object onto a ten frame as they count them and then double check their answer at the end.

GOING DEEPER

Ask children if they can find a quicker way of counting. For example, if asked to count 7 cubes from a group of 10 cubes, can they see 5 in the group of objects instantly then begin counting on until they get to 7?

KEY LANGUAGE

In lesson: count, how many, tick, circle, trace

Other language to be used by the teacher: larger group

STRUCTURES AND REPRESENTATIONS

Ten frames

RESOURCES

Mandatory: counters, multilink cubes, ten frames, colouring pencils

Optional: other real-life objects

 In the eTextbook of this lesson, you will find interactive links to a selection of teaching tools.

Quick recap

Give children some counters and cubes on their table (up to 10) and ask them to count them. Look at the different strategies they use.

Discover

Count objects from a larger group

WAYS OF WORKING Pair work

ASK

- Question ① a): *How many cubes are there in total?*
- Question ① b): *Where should we put the 4 cubes? Does it matter which 4 cubes are chosen?*

IN FOCUS In question ① b), children learn to count objects from a larger group. They need to be able to count one-by-one and ensure they move what they have counted away from the rest of the group. When counting, ensure children are saying the next number as they take one object to avoid them counting too many or too few. Some children may be able to subitise (see) 4 from the large group instantly. This should be encouraged.

PRACTICAL TIPS Children can work in pairs to place cubes or counters onto a ten frame as they count. This will help them organise their learning and give them an opportunity to check their answers.

ANSWERS

Question ① b): Children should count out 4 cubes and have 6 left.

Discover

The person who takes the last cube wins.

I rolled a 4. I need to take 4 cubes.

① a) Count out 10 cubes.

b) Count 4 cubes from the group.

20

PUPIL TEXTBOOK 1A PAGE 20

Share

WAYS OF WORKING Whole class teacher led

ASK

- Question ① a): *How many cubes are there? Are there the same amount on both rows?*
- Question ① b): *Why are some cubes bright red and some cubes faded? What is this showing? Why did we not start counting at zero?*

IN FOCUS In question ① b), children can see that only some of the cubes have been counted from the larger group of cubes. You may want children to put 4 cubes on to a ten frame to help them count and see 4.

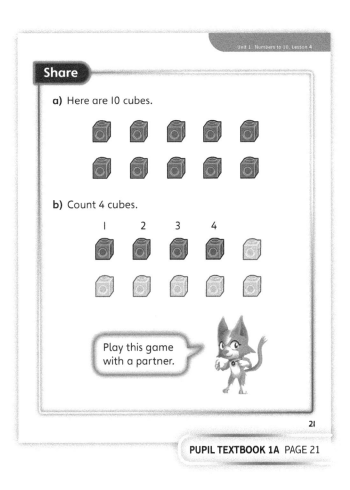

Share

a) Here are 10 cubes.

b) Count 4 cubes.

1   2   3   4

Play this game with a partner.

21

PUPIL TEXTBOOK 1A PAGE 21

Think together

WAYS OF WORKING Whole class teacher led (I do, We do, You do)

ASK

- Question **①**: *What could we use to represent the apples? Should we move each object as we count it?*

IN FOCUS In questions **①** and **②**, children count from a larger group in ones. In question **③**, children may begin to think about counting in a different way or may use their subitising skills (instantly recognising small amounts).

STRENGTHEN Use cubes or counters and ten frames to count from a larger group. Ensure children move each object as they count the next number. This will avoid them counting too many or too few. The one-to-one correspondence with the number is essential.

DEEPEN Ask children if they can subitise to count faster.

ASSESSMENT CHECKPOINT Questions **①** and **②** will give you a good indication as to whether children are confident counting from a larger group. If they are not confident, repeat more questions like these two.

ANSWERS

Question **①**: Children should stop counting with 2 apples left at the end of the line.

Question **②** a): Children should trace around a group of 2 adjacent vehicles or 2 individual vehicles.

Question **②** b): Children should trace around a group of 3 adjacent vehicles or 3 individual vehicles.

Question **②** c) Children should trace around a group of 4 adjacent vehicles or 4 individual vehicles.

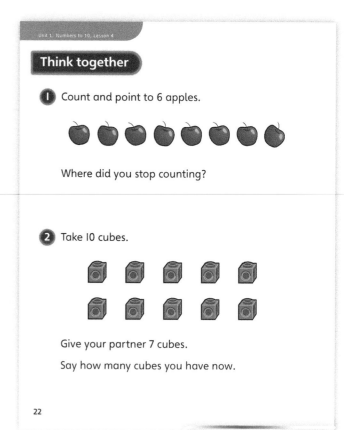

PUPIL TEXTBOOK 1A PAGE 22

PUPIL TEXTBOOK 1A PAGE 23

Practice

WAYS OF WORKING Independent working

IN FOCUS Children tick or shade in pictures to show that they can count from a larger group rather than count all objects in a group. In question ❶, children can count as they tick. In question ❷, children may find it a little more difficult to keep track as they shade in. In question ❸, children will shade 5 apples each time and this remains the same, but the total number of apples is different. Children should be able to see they are shading the same number each time.

STRENGTHEN Reinforce learning from the previous lesson by representing the pictures with objects and placing them on a ten frame. For question ❷, children could build the number on a ten frame first then remove a counter as they shade each picture until there are none left.

DEEPEN Encourage children to count in groups rather than one-by-one.

THINK DIFFERENTLY Question ❸ asks children to shade in the same number of apples three times, from a decreasing group size.

ASSESSMENT CHECKPOINT Question ❶ assesses whether children can confidently count from a larger group with no other distractions.

ANSWERS Answers for the **Practice** part of the lesson can be found in the *Power Maths* online subscription.

Reflect

WAYS OF WORKING Pair work

IN FOCUS Children practise counting a variety of numbers from a larger group.

ASSESSMENT CHECKPOINT Check children are counting accurately and organising their learning appropriately. Ask children to check each other's answers.

ANSWERS Answers for the **Reflect** part of the lesson can be found in the *Power Maths* online subscription.

After the lesson ⏸

- Are children able to count from a larger group?
- Can children organise their learning appropriately so that they do not get confused?
- Are children able to represent pictures using objects?

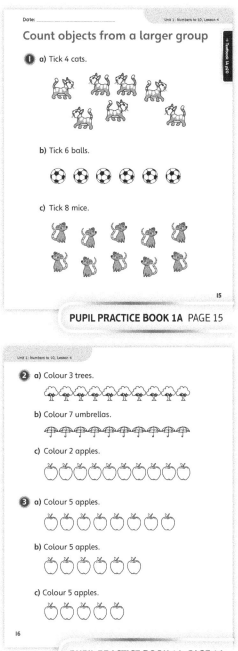

PUPIL PRACTICE BOOK 1A PAGE 15

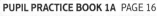

PUPIL PRACTICE BOOK 1A PAGE 16

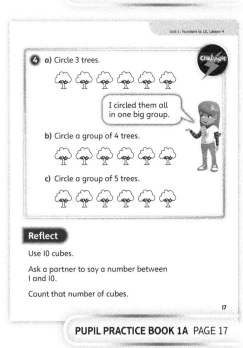

PUPIL PRACTICE BOOK 1A PAGE 17

Count on from any number

Learning focus

In this lesson, children will develop their understanding of counting by counting on from any given starting number. They will link the skill of counting concrete materials to the abstract numerals.

Before you teach

- What resources will you provide for children who find starting from a number other than 1 difficult?
- How will you provide scaffolding to build up children's confidence with counting on from any number?

NATIONAL CURRICULUM LINKS

Year 1 Number – number and place value

Count to and across 100, forwards and backwards, beginning with 0 or from 1, or from any given number.

Identify and represent numbers using objects and pictorial representations including the number line, and use the language of: equal to, more than, less than (fewer), most, least.

ASSESSING MASTERY

Children can correctly count on from any number within 10, up to and including 10. Children can relate the abstract numeral to a group of concrete objects and can recognise that adding one more increases the count by one.

COMMON MISCONCEPTIONS

Children may always begin counting from 1 rather than starting at the given number. Ask:
- *Is your starting number 1?*
- *What can you use to represent the starting number?*

Children may repeat the starting number rather than going straight to the next number. Ask:
- *What is the next number?*

STRENGTHENING UNDERSTANDING

Ask children to make their starting number using cubes or counters and build it up one at a time. Ask: *What number have you got now? So, what is the next number?* Ask them to continue this to 10. Ensure children have access to a completed number track to support them with their counting. Ask: *Do you need the whole number track? Where are you starting from? What comes next?*

GOING DEEPER

As children become more confident in counting on from any given number, they could begin to do this more abstractly and you can gradually remove scaffolding to do this. Ask: *What number is after 5? How do you know? What comes next?*

KEY LANGUAGE

In lesson: ~~count on~~ 1, 2, 3, 4, 5, 6, 7, 8, 9, 10, how many

Other language to be used by the teacher: start, end

STRUCTURES AND REPRESENTATIONS

Ten frame, number tracks

RESOURCES

Mandatory: ten frames, multilink cubes, counters

Optional: bead strings, number tracks, dice, number cards, rekenrek

 In the eTextbook of this lesson, you will find interactive links to a selection of teaching tools.

Quick recap

Give children some counters and cubes and ask them to put them on a ten frame. Can they tell you how many are on the ten frame without counting? Can they make a number on a ten frame without counting?

Discover

Count on from any number

WAYS OF WORKING Pair work

ASK
- Question ❶ a): *What number did the child start with?*
- Question ❶ a): *What number comes after the first number?*
- Question ❶ a): *Can you make the numbers?*
- Question ❶ b): *What number comes next? How do you know?*
- Question ❶ b): *Can you continue the counting?*
- Question ❶ b): *Did you need to count from 1 to know what comes next?*

IN FOCUS Question ❶ b) provides opportunity for children to continue somebody else's counting, as they begin to count on from any number.

PRACTICAL TIPS Ask a child to come up to the front and count '1, 2, 3, 4' and then the class continue the counting. Where did they start counting from?

ANSWERS

Question ❶ a): Next number is 5.

Question ❶ b): Ensure children count on from 4 as far as they can.

Discover

❶ a) What is the next number?

b) Count on from 4.

How far can you count?

24

PUPIL TEXTBOOK 1A PAGE 24

Share

WAYS OF WORKING Whole class teacher led

ASK
- Question ❶ a): *How did you use ten frames to help you? Why is one counter a different colour? What does this represent?*
- Question ❶ b): *How does the number track help?*
- *Does it matter that the number track does not start at 1?*

IN FOCUS Questions ❶ a) and ❶ b) represent counting on from 4 in different ways. Ask: *What is the same and what is different about the methods? Are they the same or different than the methods they used?* They could make the ten frame and continue adding to it to make links to the number track.

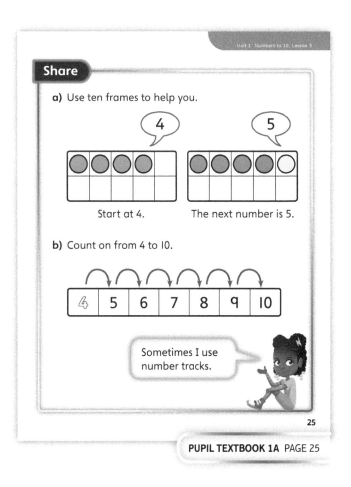

PUPIL TEXTBOOK 1A PAGE 25

Think together

WAYS OF WORKING Whole class teacher led (I do, We do, You do)

ASK

- Question ❶: *How many counters are there?*
- *If you add one more, how many counters are there now?*
- *Where does that number go on the number track?*
- *What number comes next?*

IN FOCUS Questions ❶ and ❷ provide opportunity for children to practise counting on from any number. They should make clear links between the two representations used in question ❶ and how they are similar. They could then use a similar method to support them in question ❷ if needed.

STRENGTHEN Ensure children have access to concrete materials and number tracks to support them in counting on from any number. Ensure they can identify the correct starting number when working on a number track.

DEEPEN Question ❸ is more open ended and gives children opportunity to count on from any number. They should repeat this and be prompted to spot any patterns. Ask: *Do you always count on the same amount? When do you have to count the most? When do you have to count the least? What do you notice?*

ASSESSMENT CHECKPOINT Question ❸ should be used to assess children's understanding of counting on from any number. Do they identify the correct starting number? Do they count on straight away or do they go back to 1? Do they count correctly to 10?

ANSWERS

Question ❶: 3, 4, 5, 6, 7, 8, 9, 10

Question ❷: 5, 6, 7, 8, 9, 10

Question ❸: Children choose different starting numbers to count to 10.

Unit 1: Numbers to 10, Lesson 5

Think together

❶ Count on from 3.

⬤	⬤	⬤		

1	2	3							

❷ Count on from 5.

5

26

PUPIL TEXTBOOK 1A PAGE 26

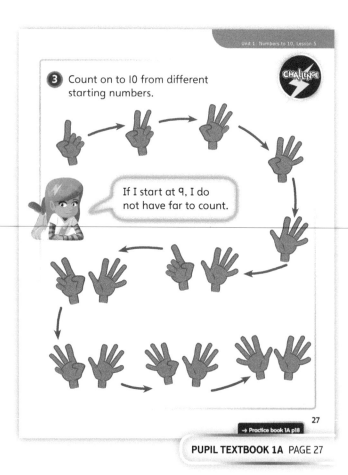

❸ Count on to 10 from different starting numbers.

CHALLENGE

If I start at 9, I do not have far to count.

27

→ Practice book 1A p18

PUPIL TEXTBOOK 1A PAGE 27

Practice

WAYS OF WORKING Independent thinking

IN FOCUS Question ❶ provides children with scaffolding to count on from any number abstractly. Question ❷ then requires children to count on from 5 to 10. Ask:
- *What number comes next?*
- *How do you know?*
- *What could you use to help you?*

STRENGTHEN Ensure children have access to concrete materials and number tracks to support them with counting on. They could count out loud to help them. If you start a count, can they continue it?

DEEPEN In question ❹, children need to interpret both the familiar horizontal number track as well as a vertical one. When they have completed it, prompt them to discuss Dexter's question. Ask: *Why does this happen? Is this the only way this could happen?*

ASSESSMENT CHECKPOINT Questions ❶ to ❸ provide opportunity to assess children's understanding of counting on from any number. Ensure they are accurately counting and not missing any numbers out or repeating numbers.

ANSWERS Answers for the **Practice** part of the lesson can be found in the *Power Maths* online subscription.

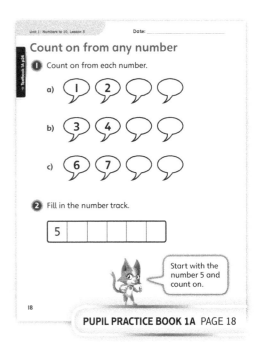

PUPIL PRACTICE BOOK 1A PAGE 18

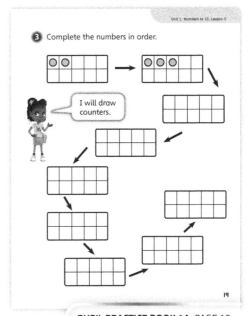

PUPIL PRACTICE BOOK 1A PAGE 19

Reflect

WAYS OF WORKING Pair work

IN FOCUS Children choose any random number between 0 and 10 and work with a partner to count on. They should now be beginning to work abstractly but could be encouraged to prove their answers or explain how they know.

ASSESSMENT CHECKPOINT By the end of this lesson, children should be able to confidently count on from any number, start a count and continue it. Can they do this confidently without always going back and starting from 1?

ANSWERS Answers for the **Reflect** part of the lesson can be found in the *Power Maths* online subscription.

After the lesson ⏸

- Are all children able to count on from any number within 10, without needing to start from 1?
- How will you use the representations used in this lesson to support later learning?
- How can you ensure children continue to practise this if they need to?

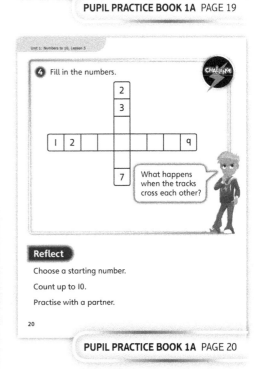

PUPIL PRACTICE BOOK 1A PAGE 20

One more

Learning focus

In this lesson, children will learn to find one more than a given number. They will investigate further the place value of numbers from 0 to 10 and consider what 'one more' means.

Before you teach

- Do children have a preferred concrete material that they always choose to use when representing amounts?
- How will you provide scaffolding for the development of children's ability to reason about place value?

NATIONAL CURRICULUM LINKS

Year 1 Number – number and place value

Given a number, identify one more and one less.

Identify and represent numbers using objects and pictorial representations including the number line, and use the language of: equal to, more than, less than (fewer), most, least.

ASSESSING MASTERY

Children can reliably and confidently count one more from any given number between 0 and 10. Children can explain what 'one more' means in terms of a number's comparative place value.

COMMON MISCONCEPTIONS

Children may be able to recite counting forwards in ones but may not be able to explain how counting forwards one more actually changes the amount. Ask:
- *Can you use the cubes to show me what happens when you count one more?*
- *How has your collection of cubes changed?*

Children may count forwards from 0 or 1 instead of beginning at the number they are finding one more than. Ask:
- *What number did you want to count forwards from?*
- *Can you show me that number in cubes?*
- *How many more cubes will you have if you count one more?*
- *Can you show me that number on a number track? Where will you be on the number track if you count one more?*

STRENGTHENING UNDERSTANDING

Make sure that all representations that have been used in the previous lessons are available. Children should be confident using ten frames by now, so use ten frames to help scaffold children's understanding of having 'one more'.

GOING DEEPER

When children are confident explaining what happens when they count one more, ask them to explain what happens when they count 2 or 3 more. Ask: *Which numbers can you count forwards 3 from without the result being greater than 10? Which numbers can you count forwards 3 from and have a result greater than 10? Why does this happen?*

KEY LANGUAGE

In lesson: one more

Other language to be used by the teacher: how many, more, greater, value, increase, prove, represent, one more than

STRUCTURES AND REPRESENTATIONS

Number track, ten frame

RESOURCES

Mandatory: multilink cubes, number track, ten frame, counters, number cards

Optional: a selection of countable real-life objects

 In the eTextbook of this lesson, you will find interactive links to a selection of teaching tools.

Quick recap

Play a game with the children where they pick a number from 0 to 9 to start with and then count on from that number to 10. Make it into a class counting game.

Discover

One more

WAYS OF WORKING Pair work

ASK

- Question **1** a): *How many dinosaurs have hatched and are out of their shells?*
- Question **1** a): *What can you use to represent the hatched dinosaurs?*
- Question **1** a): *What is going to happen with the dinosaur that is still hatching?*
- Question **1** b): *If another egg that you cannot see hatches, how many dinosaurs will you have now?*

IN FOCUS Question **1** b) requires children to visualise the increase in dinosaurs. This visualisation could be supported with toy dinosaurs or cubes to represent the dinosaurs. This is the first time that children will encounter the lesson's key language: 'one more'.

PRACTICAL TIPS Have 5 children crouch down at the front of the class with a picture of a dinosaur head in front of them. When you touch their head, they 'hatch' and stand up for children in the class to count.

ANSWERS

Question **1** a): There are 4 dinosaurs now.

Question **1** b): There are 5 dinosaurs now.

Discover

1 a) How many ?

b) One more hatches.

How many now?

28

PUPIL TEXTBOOK 1A PAGE 28

Share

WAYS OF WORKING Whole class teacher led

ASK

- Question **1** a): *How do the ten frames and blocks show that 5 is one more than 4? Can this be shown clearly as a picture?*
- Question **1** b): *If you wanted to find one more than 3, at what number should you begin counting? Why?*

IN FOCUS Question **1** b) reinforces the idea that it is not necessary to begin counting from 1 every time children want to find one more than another number. They do this in the number sentence by only showing the number that children are finding one more than, instead of showing all the numbers preceding it.

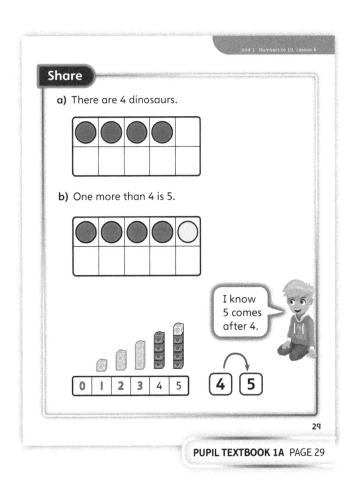

PUPIL TEXTBOOK 1A PAGE 29

Think together

Think together

WAYS OF WORKING Whole class teacher led (I do, We do, You do)

ASK

- Question ❶: *What number did you point to?*
- *Where is one more on the number track?*
- *How do you know?*
- *What could you do if you did not have the number track?*
- *How many cubes do you need to add?*

IN FOCUS Question ❶ asks children to relate the concept of one more to the position of numbers on a number track. Encourage children to repeat this multiple times and ask them what they notice about their answer. Ask: *What do you notice about the number of cubes?* In question ❷, children can then use this understanding to recognise that they need to add one more cube to find one more.

STRENGTHEN Ensure a number track is available for children who are finding the concept of finding one more challenging.

DEEPEN Encourage children to explain their answers. In question ❸, ask: *How can you help Dexter recognise what number is shown in each part? How can you find one more?*

ASSESSMENT CHECKPOINT Question ❶ provides opportunity to assess whether children can identify one more on a number track, whilst question ❷ encourages them to work without one. Ask children to find one more than a given number and explain their reasoning to see how confident they are in this.

ANSWERS

Question ❶: Children should choose any number between 0 and 10 and say one more.

Question ❷: One more than 3 is 4.

Question ❸ a): One more than 5 is 6.

Question ❸ b): One more than 9 is 10.

Question ❸ c): One more than 1 is 2.

PUPIL TEXTBOOK 1A PAGE 30

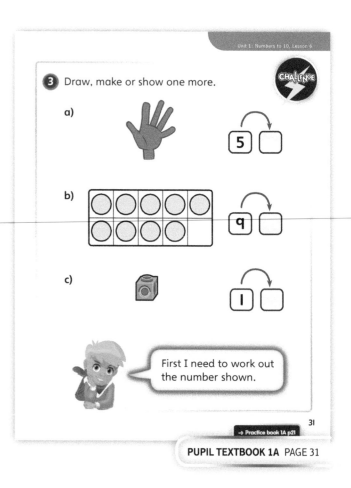

PUPIL TEXTBOOK 1A PAGE 31

Practice

WAYS OF WORKING Independent thinking

IN FOCUS Question ❶ provides visual representation to support children in finding one more. The scaffolding in the questions means that children can use their previous learning to help them by counting forwards to find the answer if needed. They should still be able to identify that their answer is one more than the previous number. In question ❷, children should recognise that each time they fill in a number on the number track they are finding one more.

STRENGTHEN Make sure a number track is available for children to refer to when answering all of the questions. They could also use a ten frame and counters to support them.

DEEPEN In question ❸, children need to identify one more without any visual representations. Ask: *What could you draw to help you to answer the questions? What else can you use?*

In question ❹, children need to recognise that sometimes the value they are given is the one more. Ask: *How can you check your answers?*

ASSESSMENT CHECKPOINT Questions ❶ to ❸ should help you to assess if children can find one more than a given number. Assess children's understanding and expose potential misconceptions by asking: *How do you know? Can you prove this? Can you show me this another way?*

ANSWERS Answers for the **Practice** part of the lesson can be found in the *Power Maths* online subscription.

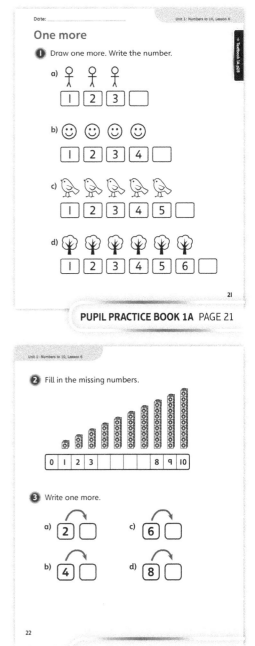

PUPIL PRACTICE BOOK 1A PAGE 21

PUPIL PRACTICE BOOK 1A PAGE 22

Reflect

WAYS OF WORKING Pair work

IN FOCUS This activity provides opportunity for children to practise finding one more. Working with a partner, they should represent different numbers and challenge each other to find one more and explain their answers.

ASSESSMENT CHECKPOINT Assess if children can confidently and consistently find one more in an efficient way. Decide if children can explain what one more means and equates to.

ANSWERS Answers for the **Reflect** part of the lesson can be found in the *Power Maths* online subscription.

PUPIL PRACTICE BOOK 1A PAGE 23

After the lesson ⏸

- Are children able to explain clearly what 'one more' means?
- Are children able to recognise why counting on from the given number is more efficient?
- Which representation best demonstrated the idea of 'one more'? What made it work so well?

69

Count backwards from 10 to 0

Learning focus

In this lesson, children use their knowledge and understanding of counting forwards to 10 to help them count backwards from 10.

Before you teach ⏸

- Are children ready to count backwards? How confident are they at counting forwards?
- How confident are children with the concept of 0?
- What real-life contexts could you use to frame this concept?

NATIONAL CURRICULUM LINKS

Year 1 Number – number and place value

Count to and across 100, forwards and backwards, beginning with 0 or 1, or from any given number.

ASSESSING MASTERY

Children can confidently count backwards to 0 from a given number up to, and including, 10. Children can use their knowledge of counting forwards and backwards to recognise patterns and complete sequences.

COMMON MISCONCEPTIONS

Children may be inclined to count forwards from a particular starting point as this may be what they are more comfortable with. Ask:
- *Which way are you counting today?*
- *Can you trace that direction on the number track with your finger?*
- *How is it different to counting forwards?*

Children may miscount and either count the same number twice or miss a number. Ask:
- *Have you counted all the numbers once?*

STRENGTHENING UNDERSTANDING

Give children number cards, each of which should show a digit, the corresponding number in words and the corresponding number in pictures. These will support understanding throughout this lesson. Display a number track prominently in the classroom for children to refer to during the lesson.

Reinforce understanding by counting backwards during daily activities such as tidying up or crossing the classroom to sit on the carpet.

GOING DEEPER

Using number cards, children turn over a card and count backwards from that number to 0. Children could then turn over two number cards, identify which number is bigger and count backwards from the bigger number to the smaller one.

KEY LANGUAGE

In lesson: pattern, count up, count back

Other language to be used by the teacher: 1, 2, 3, 4, 5, 6, 7, 8, 9, 10, how many, number, next, number track

STRUCTURES AND REPRESENTATIONS

Ten frames, number track

RESOURCES

Mandatory: ten frames, multilink cubes, number tracks, six-sided dice, number cards, counters

Optional: a selection of real-life countable objects, a cardboard tube or other item which you could use as a rocket

 In the eTextbook of this lesson, you will find interactive links to a selection of teaching tools.

Quick recap 🔁

Tell children they are going to play a game. They need 5 counters, or cubes or pencils. Ask them to count out 5 from the tray, pot or box.

Discover

Count backwards from 10 to 0

WAYS OF WORKING Pair work

ASK

- *Can you recognise a pattern in the numbers?*
- *When you count backwards, what number comes before 6?*
- *When you count backwards, what number comes after 5?*
- *How is this the same and how is this different to when you counted forwards?*

Discover

IN FOCUS Question ① a) could be supported by using a video of a rocket launch or by 'launching' a rocket that children make out of plastic bottles. The question should allow you to assess children's confidence when counting backwards.

Question ① b) provides an opportunity to discuss what number children should stop counting at. Are children confident when counting backwards to 0?

PRACTICAL TIPS Use a cardboard tube or other item as a rocket and recreate the **Discover** scenario, with the rocket taking off when children have completed the activity.

ANSWERS

Question ① a): The number 5 comes next.

Question ① b): You stop counting at 0.

① a) What number comes next?

b) How far back can you count?

Where do you stop counting?

32

PUPIL TEXTBOOK 1A PAGE 32

Share

WAYS OF WORKING Whole class teacher led

ASK

- Question ①: *Can you spot a pattern?*
- Question ① a): *What happens to the amount of blocks each time you count backwards?*
- Question ① b): *What amount is smaller than 1? Explain how you know.*
- Question ① b): *Could you show these numbers in a different way?*
- Question ① b): *How did you represent the numbers in the last lesson? Could you use that method again?*
- Question ① b): *How will you know what comes next?*

IN FOCUS Questions ① a) and ① b) support children's understanding of counting backwards and demonstrate how the amount decreases as they count backwards. The second representation also supports children in their understanding of 0 equalling nothing and what this looks like in comparison to the other amounts.

Share

a)

| 10 | 9 | 8 | 7 | 6 | 5 |

The number 5 comes next.
Say the next numbers.

Is there a pattern?

This number track looks different.

b)

| 10 | 9 | 8 | 7 | 6 | 5 | 4 | 3 | 2 | 1 | 0 |

You stop counting at 0.

33

PUPIL TEXTBOOK 1A PAGE 33

Think together

WAYS OF WORKING Whole class teacher led (I do, We do, You do)

ASK

• *What number comes next? How can you check?*
• *What did you use to help you count?*
• *How many blocks do you remove each time you count backwards?*
• *What can you do to help you count backwards from any number?*
• *Can you show your counting by drawing a picture?*

IN FOCUS Questions ❶ and ❷ support children's understanding of counting backwards and demonstrates how the amount decreases as they count backwards. Question ❷ also supports children in their understanding of 0 equalling nothing and what this looks like in comparison to the other amounts.

STRENGTHEN Have a number track available for children who are finding the process of counting backwards more challenging. Ask children to make the numbers as towers of cubes and get them to take one cube off each time - relating this to the number track as they count.

DEEPEN In question ❸, ask: *What is the same and what is different about counting forwards and counting backwards? Can you complete both number tracks the other way? What do you notice?*

ASSESSMENT CHECKPOINT Questions ❶ to ❸ should help you assess children's ability to recognise numerals and count backwards from them. Can children explain how they know what number comes next? Can they show you how they know?

ANSWERS

Question ❶: Next numbers are 4, 3.

Question ❷: Next numbers are 2, 1, 0.

Question ❸ a): 4, 5, 6, 7, 8, 9, 10

Question ❸ b): 8, 7, 6, 5, 4, 3, 2, 1, 0

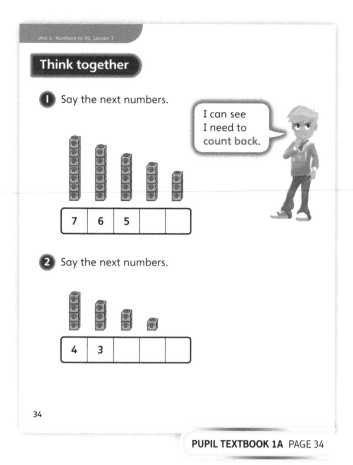

PUPIL TEXTBOOK 1A PAGE 34

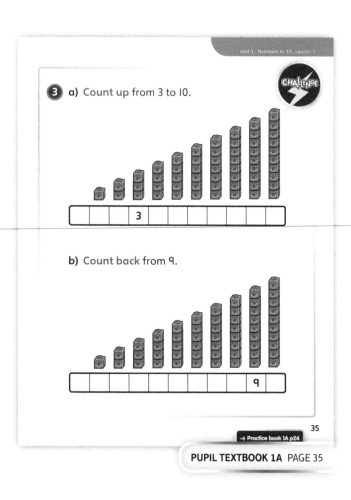

PUPIL TEXTBOOK 1A PAGE 35

Practice

WAYS OF WORKING Independent thinking

IN FOCUS Questions ❶ to ❸ allow children to practise the skill of counting backwards from a given number. The questions are increasingly abstract, gradually requiring children to rely more on their own knowledge of numbers in order to solve them.

STRENGTHEN Provide a number track for children to refer to. Children could also have access to a ten frame and counters to build the numbers and take one counter away each time they count backwards.

DEEPEN Questions ❹ and ❺ challenge children to recognise whether a number sequence is counting forwards or backwards and complete it appropriately.

THINK DIFFERENTLY Question ❹ asks children to think differently by leaving the starting numbers blank in questions ❹ a), b) and d), requiring children to work inversely to the pattern to solve them.

ASSESSMENT CHECKPOINT Questions ❶ to ❸ should help you assess children's ability to count backwards from a given number and represent the sequences in different ways. Check that children are able to correctly represent the patterns in all forms of the numbers (as pictures, digits and written names).

ANSWERS Answers for the **Practice** part of the lesson can be found in the *Power Maths* online subscription.

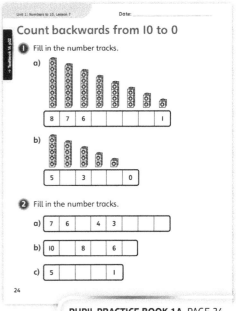

PUPIL PRACTICE BOOK 1A PAGE 24

PUPIL PRACTICE BOOK 1A PAGE 25

Reflect

WAYS OF WORKING Independent thinking

IN FOCUS At this point, children should be able to roll a random number below 10 and count forwards and backwards from it accurately and confidently. Ask children where they should put their number on the number track so that the 10 and 1 appear at the beginning and end of their number track. Can they explain their ideas?

ASSESSMENT CHECKPOINT This **Reflect** activity should help you assess if children are counting forwards and backwards confidently and fluently. Are there any numbers missing from their sequences?

ANSWERS Answers for the **Reflect** part of the lesson can be found in the *Power Maths* online subscription.

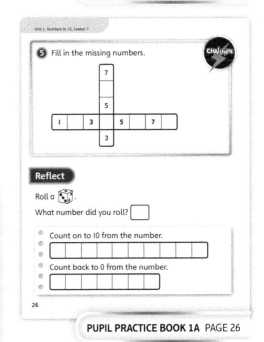

PUPIL PRACTICE BOOK 1A PAGE 26

After the lesson ⏸

- Do any children still have any misconceptions about counting backwards?
- Are children sufficiently confident and able to explain and show the differences between counting forwards and backwards?
- Is there anything that you will do differently when you approach this concept again in the future?

One less

Learning focus

In this lesson, children will learn to find one less than a given number. They will investigate further the place value of numbers from 0 to 10 and consider what 'one less' means.

Before you teach

- In Lesson 7, how did children respond to the questions that required them to think inversely?
- How could you use this as a starting point for Lesson 8?

NATIONAL CURRICULUM LINKS

Year 1 Number – number and place value

Given a number, identify one more and one less.

ASSESSING MASTERY

Children can reliably and confidently count one less from any given number between 1 and 10. Children can explain what 'one less' means in terms of a number's comparative place value.

COMMON MISCONCEPTIONS

Children may be able to recite counting backwards in ones but may not be able to explain how counting one less actually changes the amount. Ask:
- *Can you use cubes to show me what happens when you count one less?*
- *How has your collection of cubes changed?*

Children may count forwards instead of backwards. Ask:
- *On a number track, can you show me which way you would go to count backwards?*

STRENGTHENING UNDERSTANDING

Make sure that all representations that have been used in the previous lessons are available. Children should be confident using ten frames by now, so use ten frames to help scaffold children's understanding of having 'one less'.

Use the concept of musical chairs (as shown in **Discover**). Children could play musical chairs with toys, discussing how many chairs will be left and how many toys would be able to sit down as each chair is taken away.

GOING DEEPER

When children are confident explaining what happens when they count backwards by 1, ask them to explain what happens when they count backwards by 2 or 3. Ask: *Which numbers can you count backwards 3 from without the result being smaller than 0?*

KEY LANGUAGE

In lesson: one less, one less than

Other language to be used by the teacher: how many, another, pattern, fewer, value, less, decrease

STRUCTURES AND REPRESENTATIONS

Number track, ten frame

RESOURCES

Mandatory: multilink cubes, number tracks, ten frames, counters

Optional: a selection of countable real-life objects

 In the eTextbook of this lesson, you will find interactive links to a selection of teaching tools.

Quick recap

Play a counting game or sing a counting song where the children count forwards and backwards from 1 to 10.

Discover

WAYS OF WORKING Pair work

ASK

What is happening in the picture?
- Question ①: *What is the teacher doing?*
- Question ①: *If a chair is taken away, can more children sit down or can fewer children sit down?*
- Question ① b): *If 1 chair is taken away, how many are left?*
- Question ① b): *Can every child in the picture sit down?*
- Question ① b): *Why is the number of children always one more than the number of chairs?*

IN FOCUS Questions ① a) and ① b) could be approached by playing musical chairs in an appropriate space. Children are likely to recognise the game of musical chairs, which provides a good introduction to the concept of 'one less'. Discuss with children how they might represent the chairs with towers, cubes or counters on a ten frame.

PRACTICAL TIPS Play musical chairs with 7 children in the classroom to recreate the **Discover** scenario.

ANSWERS

Question ① a): There are 7 chairs.

Question ① b): There are 6 chairs now.

Unit 1: Numbers to 10, Lesson 8

One less

Discover

① a) How many chairs?

  b) One chair is removed.
     How many chairs now?

36

PUPIL TEXTBOOK 1A PAGE 36

Share

WAYS OF WORKING Whole class teacher led

ASK

- Question ① a): *How can you represent the number of chairs?*
- Question ① a): *How can you use cubes or counters to show that the number of chairs is one less?*
- Question ① b): *Can you have one less than 1? Can you prove it with resources or in a picture?*

IN FOCUS Question ① b) introduces children to the lesson's key language: 'one less than'. This provides an opportunity to link the concept of 'one less' to their knowledge and understanding of 'one more'. Discussing how it is similar but different should help to clarify the key differences between 'one more' and 'one less'. It should also help children to avoid counting forwards when asked to count backwards.

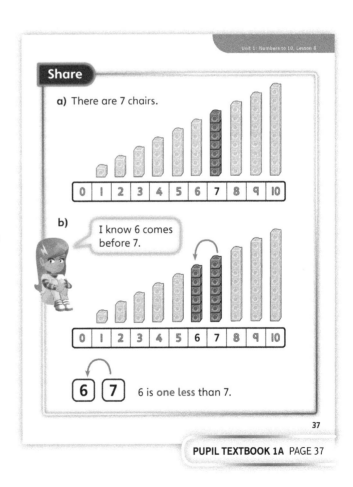

Share

a) There are 7 chairs.

| 0 | 1 | 2 | 3 | 4 | 5 | 6 | 7 | 8 | 9 | 10 |

b)

I know 6 comes before 7.

| 0 | 1 | 2 | 3 | 4 | 5 | 6 | 7 | 8 | 9 | 10 |

6   7   6 is one less than 7.

37

PUPIL TEXTBOOK 1A PAGE 37

Think together

WAYS OF WORKING Whole class teacher led (I do, We do, You do)

ASK

- Question ①: *What number did you point to?*
- Question ①: *Where is one less on the number track?*
- Question ②: *Can you show it another way?*
- Question ③: *How is finding one less similar and different to finding one more? Explain your ideas.*

IN FOCUS Questions ② and ③ support children in understanding what it means to find one less, so they do not think it only means the position of the number on the number track. Discuss the characters' ideas to ensure children understand this.

STRENGTHEN Consider linking the concept of 'one less' to real-life experiences of children and support with the representations used in the previous lessons. For example, ask: *If you have 3 sweets and I take 1 from you, will you have more or fewer sweets? How many sweets do you have?* Children could arrange the 3 sweets on a ten frame and remove one. Ask: *How many sweets were there? How many sweets are there now?*

DEEPEN Ask: *Do you notice any patterns in the numbers. Can you find one less and then find one less again? How many less have you found?*

ASSESSMENT CHECKPOINT Question ① assesses whether children can identify one less from the position on a number track. Questions ② and ③ provide opportunities to ensure children understand what is meant by one less and are able to physically find it.

ANSWERS

Question ①: Children choose different starting numbers to say one less.

Question ②: One less than 3 is 2.

Question ③ a): One less than 6 is 5.

Question ③ b): One less than 5 is 4.

Question ③ c): One less than 1 is 0.

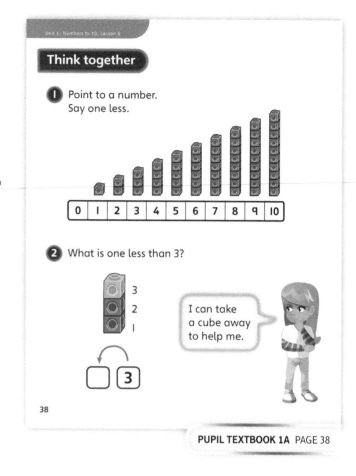

PUPIL TEXTBOOK 1A PAGE 38

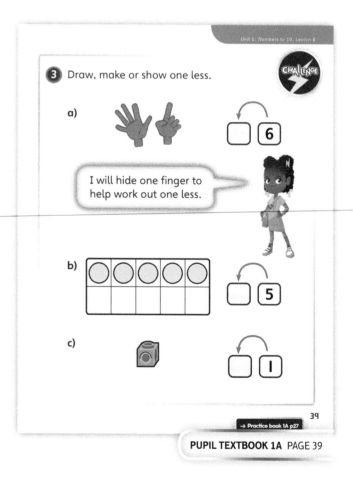

PUPIL TEXTBOOK 1A PAGE 39

Practice

WAYS OF WORKING Independent thinking

IN FOCUS Question ❶ provides visual representations to support children in finding one less. The scaffolding in the questions means that children can use their previous learning to help them by counting forwards to find the answer, if needed. They should still be able to identify that their answer is one less than the end number.

STRENGTHEN Make sure the number track from question ❷ is available for children to refer to when answering all of the questions in this section. They could also use a ten frame and counters for support.

DEEPEN In question ❸, children need to identify one less without any visual representations. Ask: *What could you draw to help you answer the questions? What else could you use?*

THINK DIFFERENTLY In question ❹, children need to decide whether the number they are finding is one more or one less. Ask: *How can you check your answers?*

ASSESSMENT CHECKPOINT Questions ❶, ❷ and ❸ should help you to assess if children can find one less than a given number. Assess children's understanding and expose potential misconceptions by asking: *How do you know? Can you prove this? Can you show me this another way?*

ANSWERS Answers for the Practice part of the lesson can be found in the *Power Maths* online subscription.

Reflect

WAYS OF WORKING Pair work

IN FOCUS Discuss how this lesson's learning is similar and linked to what children learnt in Lesson 6. Can children use their knowledge of one more to help them find one less?

ASSESSMENT CHECKPOINT This **Reflect** activity should help you assess if children can confidently and consistently find one less in an efficient way. Assess if children can explain what 'one less' means and equates to. Are children confident discussing the links between finding one more and one less? Assess if children can justify and prove their ideas through confident use of different representations of the numbers.

ANSWERS Answers for the **Reflect** part of the lesson can be found in the *Power Maths* online subscription.

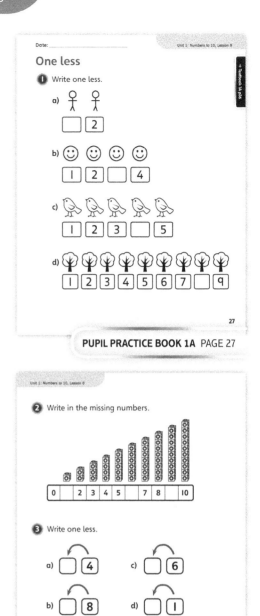

PUPIL PRACTICE BOOK 1A PAGE 27

PUPIL PRACTICE BOOK 1A PAGE 28

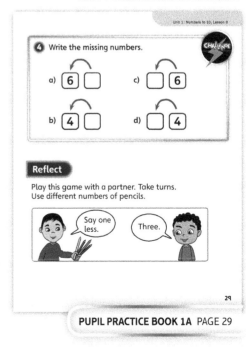

PUPIL PRACTICE BOOK 1A PAGE 29

After the lesson ⏸

- Do children clearly understand the links between the topics they have learnt about in Lessons 6 and 8?
- Do children clearly understand the links between 'one more' and 'one less'?
- Are children more confident with 'one more' than they are with 'one less' or vice versa? How will you support them in the future so that they become equally confident with both?

Compare groups

Learning focus

In this lesson, children will compare groups of objects. Children will identify, when given two groups of objects, whether one group has more objects than the other.

Before you teach

- Are children confident counting forwards and backwards?
- What resources and representations will you make available from previous lessons to support their learning?

NATIONAL CURRICULUM LINKS

Year 1 Number – number and place value

Identify and represent numbers using objects and pictorial representations including the number line, and use the language of: equal to, more than, less than (fewer), most, least.

ASSESSING MASTERY

Children can match one object in each group using one-to-one correspondence (by drawing lines or lining up) to determine if there are enough or the same number, or an unequal number of objects in each group.

COMMON MISCONCEPTIONS

Children may not correctly do one-to-one correspondence from one object to another object in each group, particularly if they are trying to trace with their finger. Encourage children to draw a line from one object to another. Check they have only drawn one line from and to each object.

Children may instead line up one object in front of another to show the one-to-one correspondence.

STRENGTHENING UNDERSTANDING

To strengthen understanding of one-to-one correspondence, children draw lines from one object in one group to one object in the other group. Check when they line up the objects that they only line up one object in front of another. Ask: *Do you have only one line coming from each object? Which objects are there more of and which are there fewer of? How do you know?*

Prepare pictorial representations of the objects in the **Textbook** for children to manipulate, group and count when answering the questions.

GOING DEEPER

Some children will begin to go deeper by realising that they can work out how many fewer and more by counting the number of objects in each group. Ask: *How many more of this object do you have than of that object? How do you know?*

KEY LANGUAGE

In lesson: compare

Other language to be used by the teacher: match, sorted, count, equal, less than, fewer than, greater than, more than, share

RESOURCES

Mandatory: multilink cubes, 2D shapes (squares and circles), ten frame, counters

Optional: number tracks, a selection of countable real-life objects

 In the eTextbook of this lesson, you will find interactive links to a selection of teaching tools.

Quick recap

Ask children to show you a certain number of fingers or cubes. Start with numbers from 1 to 5 and then gradually increase to numbers up to 10. Mix them up so they are not in order.

Discover

WAYS OF WORKING Pair work

ASK

- Question ① a): *How many children are there? How many balls are there?*
- Question ① b): *Can each child have a ball? How do you know?*

IN FOCUS Question ① a) prompts a discussion about the methods children could use to help them compare groups of objects.

PRACTICAL TIPS Use 6 children and 6 balls to recreate the **Discover** activity in the classroom. Ask them each to get a ball. Can they each have one?

ANSWERS

Question ① b): Yes, each child can have a ball.

Compare groups

Discover

① **a)** Use your finger to draw a line from each child to a ball.

b) Can each child have a ball?

40

PUPIL TEXTBOOK 1A PAGE 40

Share

WAYS OF WORKING Whole class teacher led

ASK

- Question ① b): *How is the matching made clear?*
- Question ① b): *Did anyone compare the objects in a different way?*

IN FOCUS Question ① scaffolds comparing groups for children. Ask: *What method did Dexter use? How did lining up each child with a ball help Dexter show that each child can have a ball?*

Share

a) Draw a line from each child to a ball.

b) Each child can have a ball.

I lined up each child with a ball.

41

PUPIL TEXTBOOK 1A PAGE 41

79

Think together

WAYS OF WORKING Whole class teacher led (I do, We do, You do)

ASK

- Question ❶: *What can you do to decide if there are enough apples?*
- Question ❷: *What do you predict? What can you do to check?*
- Question ❷: *How many more do you need so each child can have one?*

IN FOCUS Questions ❶ and ❷ ensure children can use matching to compare groups. Make sure they only match to one thing and do not draw more than one line to the same object.

STRENGTHEN Represent the people and objects in the different groups using two different coloured towers of multilink cubes. Children could count out multilink cubes to represent people and objects and then compare the two groups. Ask: *What are there more of? Are there enough? How can you tell?*

DEEPEN In question ❸, ask children to use Ash's method to access the question practically. Have children work through it this way and then think about how they could do it another way.

ASSESSMENT CHECKPOINT Questions ❶ and ❷ will show whether or not children can compare groups by matching. Question ❸ should help you to decide if children can organise and count groups of objects to compare. Ask children if they can organise the groups to facilitate easy comparison.

ANSWERS

Question ❶: Yes, each child can have an apple.

Question ❷: No, each child cannot have a pizza.

Question ❸: Each child can have a counter but each child cannot have a brick.

Think together

❶ Can each child have an apple?

❷ Can each child have a pizza?

42

PUPIL TEXTBOOK 1A PAGE 42

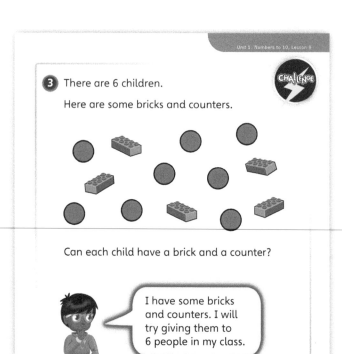

❸ There are 6 children.

Here are some bricks and counters.

CHALLENGE

Can each child have a brick and a counter?

I have some bricks and counters. I will try giving them to 6 people in my class.

43

→ Practice book 1A p30

PUPIL TEXTBOOK 1A PAGE 43

Practice

WAYS OF WORKING Independent thinking

IN FOCUS Questions ❶ and ❷ provide opportunity for children to compare groups and decide if there are enough presents/pieces of cheese. In question ❸ children draw the worm for each bird and then recognise that this means each bird can have a worm. Question ❹ uses more mugs than dishes and children could use Sparks's advice to support them.

STRENGTHEN Provide children with two different colours of multilink cubes to help them represent the questions. Ask: *What size or colour will you use to represent each item? How many of each are there? Are there enough? How do you know?*

DEEPEN In question ❺ the pictures are organised in a less structured way, which can make it more challenging for children to draw lines. Ask: *Can you use your objects to represent it more efficiently? What do you notice about your answers to questions ❺ a) and ❺ b)?* Prompt discussion about how these two answers are linked.

ASSESSMENT CHECKPOINT Questions ❶ to ❹ will help you to assess whether or not children can compare groups by matching.

ANSWERS Answers for the **Practice** part of the lesson can be found in the *Power Maths* online subscription.

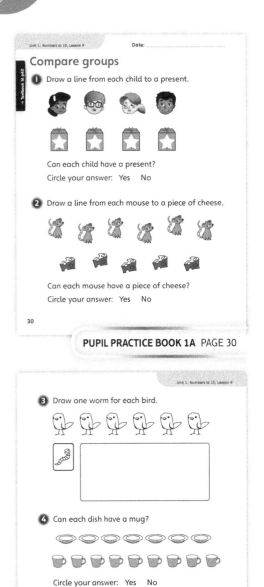

PUPIL PRACTICE BOOK 1A PAGE 30

PUPIL PRACTICE BOOK 1A PAGE 31

Reflect

WAYS OF WORKING Independent thinking

IN FOCUS This activity develops children's ability to compare groups by matching. Encourage children to reason and justify whether there are enough counters/pencils.

ASSESSMENT CHECKPOINT By the end of this lesson, children should be able to compare groups by matching. They should be able to answer the questions: Are there enough? Are there the same number?

ANSWERS Answers for the **Reflect** part of the lesson can be found in the *Power Maths* online subscription.

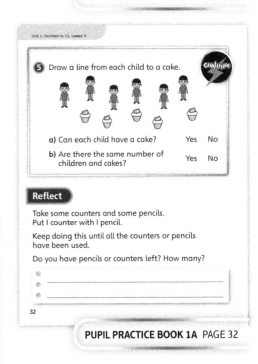

PUPIL PRACTICE BOOK 1A PAGE 32

After the lesson ⏸

- Are children able to match objects in different groups and use their matching to compare groups?
- Can children use their matching to decide whether there are enough of each object?
- Are all children able to organise groups clearly and accurately?

Fewer or more?

Learning focus

In this lesson, children will compare groups of objects. Children will identify, when given two groups of objects, whether one group has more objects than the other.

Before you teach

- Are children confident counting forwards and backwards?
- What resources and representations will you make available from previous lessons to support their learning?

NATIONAL CURRICULUM LINKS

Year 1 Number – number and place value

Identify and represent numbers using objects and pictorial representations including the number line, and use the language of: equal to, more than, less than (fewer), most, least.

ASSESSING MASTERY

Children can recognise two groups of objects, count them and explain which group has more and which group has fewer. Children can justify their reasoning using concrete, pictorial and abstract representations.

COMMON MISCONCEPTIONS

Children may miscount the objects and so compare the groups inaccurately. Ask:
- *What could you use to help you count the objects?*

Children may say that a greater number is smaller or vice versa. Ask:
- *What could you use to help you compare?*
- *Could you show these numbers using multilink cubes?*

STRENGTHENING UNDERSTANDING

To strengthen understanding of counting and comparing, discuss the different representations that children could use. Ask: *Could you use a ten frame to help you visualise the amount? Could you use multilink cubes to help you compare? How could the number track help?*

Prepare pictorial representations of the flags, sandcastles, buckets and spades in the **Textbook** for children to manipulate, group and count when answering the questions.

GOING DEEPER

Ask children to investigate by how much a group is bigger or smaller than another group. Ask: *How many more of this object do you have than of that object? How do you know?*

KEY LANGUAGE

In lesson: fewer, matched, more

Other language to be used by the teacher: match, sorted, compare, count, equal, less than, fewer than, greater than, more than

STRUCTURES AND REPRESENTATIONS

Ten frame

RESOURCES

Mandatory: multilink cubes, counters

Optional: number tracks, 2D shapes (squares and circles), ten frame, a selection of countable real-life objects

 In the eTextbook of this lesson, you will find interactive links to a selection of teaching tools.

Quick recap

Ask children to choose a number from 1 to 10 and make the number on a ten frame. They should then show a partner and agree which has the most and the fewest counters.

Discover

Unit 1: Numbers to 10, Lesson 10

Fewer or more

Discover

WAYS OF WORKING Pair work

ASK

- Question ① a): *Has each child made a sandcastle? How do you know?*
- Question ① b): *How many flags are standing up? Are there more flags or fewer flags lying down?*

IN FOCUS Question ① a) prompts a discussion about the different representations and resources that children could use to help them compare groups of objects.

PRACTICAL TIPS Replicate the **Discover** scenario in the classroom. Use photos or real-life objects to replicate the flags, sandcastles and buckets and ask 7 children to come to the front of the class.

ANSWERS

Question ① a): There are more flags.

Question ① b): There are fewer buckets.

① **a)** Are there more flags or more 🏰 ?

b) Are there fewer children or 🪣?

44

PUPIL TEXTBOOK 1A PAGE 44

Share

WAYS OF WORKING Whole class teacher led

ASK

- Question ① a): *How is the matching made clear?*
- *Did anyone compare the objects in a different way?*
- Question ① b): *How many more buckets do you need so that everyone can have a bucket?*

IN FOCUS Question ① a) introduces children to the lesson's key language: 'matched'. Discuss the meaning of 'matched' and how it has helped Dexter decide whether there are more sandcastles or more flags.

Question ① b) introduces children to the lesson's key language: 'fewer'. Discuss what fewer means, in relation to the number of people and the number of buckets.

Think together

WAYS OF WORKING Whole class teacher led (I do, We do, You do)

ASK

- Question ❶ *What do you do with the buckets? What do you do with the sandcastles? How can you show where they match?*
- Question ❸ *What shape do you predict there are more of? Explain your idea. How could you make these shapes easier to count?*

IN FOCUS Question ❶ provides scaffolding to help children compare the buckets and sandcastles and explain which group has more. In question ❷, children will use 'fewer' in their explanation and conclusion.

STRENGTHEN Represent the people and objects in the different groups using two different-coloured towers of multilink cubes. Children could count out multilink cubes to represent people and objects and then compare the two groups. Ask: *Which has more? Which has fewer? How can you tell?*

DEEPEN Use Astrid's and Flo's comments in question ❸ to discuss which method is the most efficient and effective. Ask children to try both methods and decide which one they prefer. If two children think differently, get them to work together in pairs to convince each other.

ASSESSMENT CHECKPOINT Question ❸ should help you to decide if children can organise and count groups of objects. Assess if children can organise the groups to facilitate easy comparison and explain which group has more and which group has fewer.

ANSWERS

Question ❶: There are more sandcastles.

Question ❷: There are fewer people.

Question ❸: There are more squares.

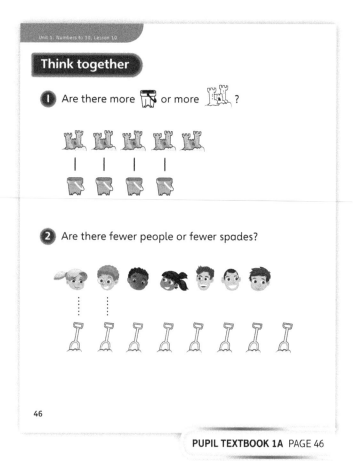

Think together

❶ Are there more 🪣 or more 🏰 ?

❷ Are there fewer people or fewer spades?

46

PUPIL TEXTBOOK 1A PAGE 46

❸ Are there more ◯ or more ☐ ? **CHALLENGE**

Use different colour counters for the ☐ and ◯ .

I will sort them first.

I will put them in two lines.

47

→ Practice book 1A p33

PUPIL TEXTBOOK 1A PAGE 47

Practice

WAYS OF WORKING Independent thinking

IN FOCUS Questions ❶ and ❸ support comparisons by arranging the objects in neat rows. The objects in questions ❷ and ❹ are not arranged as neatly and are set out so they start and end at the same points, which may lead children to think there are the same number. Discuss with children how they could make the correspondence between the apples and the bananas, and the mice and the cheese, clearer to help them decide what there are more of.

STRENGTHEN Provide children with two different colours of multilink cubes to represent the objects in the questions. When the objects are presented in a more scattered way, how can they use the cubes to arrange them more neatly? How does this help to decide what there are more of?

DEEPEN In question ❺, the objects are mixed up and not in their own rows. Ask: *How can you use your learning from the previous session to help you?* If they use Astrid's advice and match the shapes, ask: *What shapes are left over? What does this mean? Can you get to the answer a different way?*

ASSESSMENT CHECKPOINT Questions ❶ to ❹ should help you to decide whether children can compare groups using the language of more and fewer.

ANSWERS Answers for the **Practice** part of the lesson can be found in the *Power Maths* online subscription.

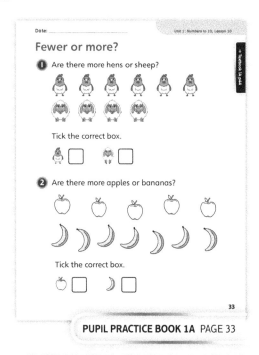

PUPIL PRACTICE BOOK 1A PAGE 33

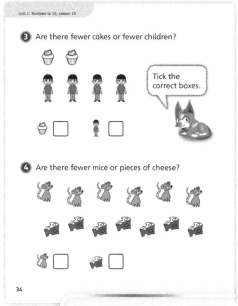

PUPIL PRACTICE BOOK 1A PAGE 34

Reflect

WAYS OF WORKING Independent thinking

IN FOCUS This **Reflect** activity develops children's ability to reason and justify. Are children able to make the connection between fewer and less than?

ASSESSMENT CHECKPOINT Can children find more than one solution? Can children identify all the amounts that are less than 5? Assess whether children are able to use a systematic approach to finding all the solutions.

ANSWERS Answers for the **Reflect** part of the lesson can be found in the *Power Maths* online subscription.

PUPIL PRACTICE BOOK 1A PAGE 35

After the lesson ⏸

- Are children able to confidently compare two groups?
- Can children use the language learnt in the lesson to describe the comparisons?
- Are all children able to organise groups clearly and accurately?

<, > or =

Learning focus

In this lesson, children will use the <, > and = signs to compare two groups of objects. They will explain their comparisons using the correct mathematical language.

Before you teach ⏸

- Are children confident in their ability to recognise numbers that are bigger or smaller than one another?
- How will you make the difference between < and > as clear as possible?

NATIONAL CURRICULUM LINKS

Year 1 Number – number and place value

Identify and represent numbers using objects and pictorial representations including the number line, and use the language of: equal to, more than, less than (fewer), most, least.

ASSESSING MASTERY

Children can describe a number as greater than, less than or equal to another number and clearly explain how the <, > and = signs are used. Children can demonstrate this knowledge using different representations.

COMMON MISCONCEPTIONS

Children are likely to confuse > (greater than) and < (less than). Ask:
- *Could you use multilink cubes to help you compare?*
- *How could the number track help you compare?*

STRENGTHENING UNDERSTANDING

To strengthen children's ability to count and compare, provide large versions of the <, > and = signs. Ask: *Can you use these signs to help you visualise the comparison?*

Encourage children to arrange different-sized towers of multilink cubes either side of each sign. Ask: *Can you point to each amount and tell me the sentence you would say to compare them? What side do you start reading from? Does the sign begin with the bigger end or the smaller end?*

GOING DEEPER

Give children a number of multilink cubes. Encourage children to use them to investigate how many different ways they can partition a given number and then make comparisons between the different parts of the number using the signs introduced in this lesson. Ask: *How many ways can you partition this number? Have you found all the comparisons you can make? Is there a pattern?*

KEY LANGUAGE

In lesson: greater than, >, fewer than, <, equal to, =

Other language to be used by the teacher: same, compare

STRUCTURES AND REPRESENTATIONS

Ten frame

RESOURCES

Mandatory: multilink cubes, counters

Optional: a selection of countable real-life objects such as tennis balls and footballs, large printed versions of the <, > and = signs for children to hold and manipulate, number track

 In the eTextbook of this lesson, you will find interactive links to a selection of teaching tools.

Quick recap

Ask children to choose a number from 1 to 10 and to represent this number on a ten frame. They should then show a partner. Then, announce the winner is the child with the most or fewest counters.

Discover

<, > or =

WAYS OF WORKING Pair work

ASK

- Question ① b): *How many multilink cubes does Ola have?*
- Question ① b): *How else could you show the amounts in this picture?*
- Question ① b): *Can you order all the amounts from most to least?*

IN FOCUS Question ① a) and ① b) ask children to compare two amounts. They move from comparing 'one more' and 'one less' to any amount more and less. Introduce the lesson's key language of comparison: 'less than' and 'greater than'.

PRACTICAL TIPS Replicate the **Discover** scenario using multilink cubes at the front of the class. Alternatively, you could give children their own multilink cubes and ask them to work in small groups to recreate the problem.

ANSWERS

Question ① a): Tim has more cubes than Lou.
Lou has fewer cubes than Tim.

Question ① b): Tim and Ola have an equal number of cubes.

Discover

Tim   Lou   Ola

① **a)** Who has more cubes, Tim or Lou?

Who has fewer?

b) Who has more cubes, Tim or Ola?

48

PUPIL TEXTBOOK 1A PAGE 48

Share

WAYS OF WORKING Whole class teacher led

ASK

- Question ① a): *What could you use the <, > and = signs to compare?*
- Question ① a): *How do the pictures show you how to use the signs?*
- Question ① a): *How are > and < the same and how are they different?*
- Question ① b): *Can you show another two numbers that are equal?*

IN FOCUS Question ① a) and ① b) introduce children to the lesson's key language. Use large printed versions of the <, > and = signs for children to manipulate and arrange objects on.

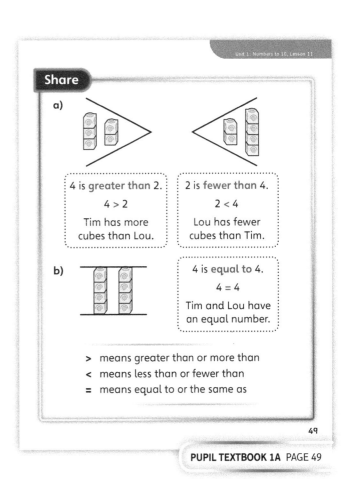

Share

a)

4 is greater than 2.
4 > 2
Tim has more cubes than Lou.

2 is fewer than 4.
2 < 4
Lou has fewer cubes than Tim.

b)

4 is equal to 4.
4 = 4
Tim and Lou have an equal number.

> means greater than or more than
< means less than or fewer than
= means equal to or the same as

49

PUPIL TEXTBOOK 1A PAGE 49

Think together

WAYS OF WORKING Whole class teacher led (I do, We do, You do)

ASK
- *How does the tower size help you compare the numbers?*
- *Which sign do you need to use?*
- *How do you know?*

IN FOCUS In questions ❶ and ❷, towers are shown to children so that they can compare their heights to support them in comparing numbers. Children could use cubes on a whiteboard and draw the correct sign around them to help them.

STRENGTHEN Ensure children have the visual representation of the signs and cubes to support them in deciding which sign they need to use.

DEEPEN In question ❸ a) children are given the correct sign in the statement and are asked to use concrete materials to prove each statement. Encourage children to explain their answers and how the height of the towers supports the comparisons. In question ❸ b), encourage children to find all the possible answers. Ask: *How do you know you have found all the answers?*

ASSESSMENT CHECKPOINT Questions ❶ and ❷ should help you assess whether children can use signs to compare numbers.

ANSWERS

Question ❶: 6 > 5

Question ❷: 3 < 5

Question ❸ b): The missing number could be 3, 2, 1 or 0.

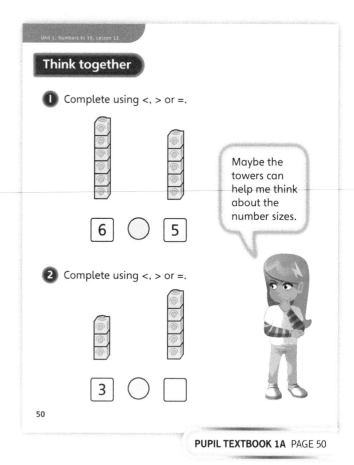

PUPIL TEXTBOOK 1A PAGE 50

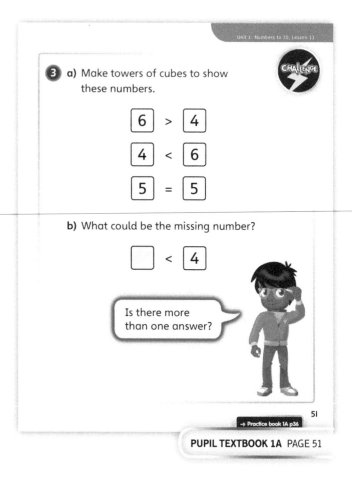

PUPIL TEXTBOOK 1A PAGE 51

Practice

WAYS OF WORKING Independent thinking

IN FOCUS Questions ❶ and ❷ require children to count and compare numbers of cubes and choose the correct inequality sign. Question ❸ requires children to draw towers of cubes to prove number sentences and support their understanding of the inequality signs.

STRENGTHEN Ensure children have access to visual representations of the inequality signs to support them in deciding which sign they need to use.

DEEPEN In question ❹, children need to think more carefully about their answers. In question ❹ a) only one set of cubes is arranged as a tower so they cannot simply compare height. Ask: *What else can you do? Can you build the towers? Do you need to build the towers?* In question ❹ b) two different sizes of cubes are used so, whilst the towers are of similar heights, this does not mean the two numbers are equal.

ASSESSMENT CHECKPOINT Questions ❶ and ❷ should help you to assess whether or not children can compare numbers using the inequality signs.

ANSWERS Answers for the **Practice** part of the lesson can be found in the *Power Maths* online subscription.

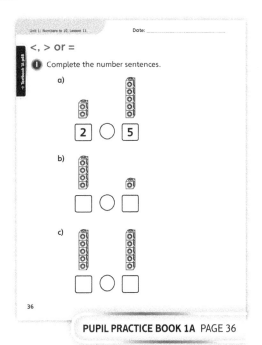

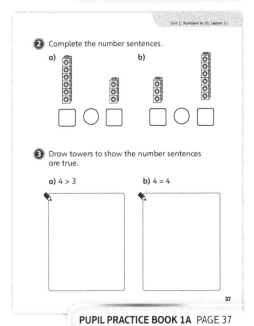

Reflect

WAYS OF WORKING Independent thinking

IN FOCUS This activity asks children to explain and demonstrate in their own way the meaning of each inequality sign. Encourage them to explain what they have done and how it helps them to remember what each sign means.

ASSESSMENT CHECKPOINT By the end of this lesson, children should be able to compare numbers using the inequality signs. Assess whether the examples they have chosen in their reflect activity are used with the correct signs

ANSWERS Answers for the **Reflect** part of the lesson can be found in the *Power Maths* online subscription.

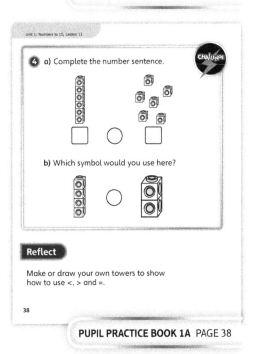

After the lesson

• Are children using the <, > and = signs accurately?
• Are children ready to use the <, > and = signs in other contexts?

Compare numbers

Learning focus

In this lesson, children will compare more abstract numbers where they are not given countable objects. Children will choose the best representation or resource to help them compare.

Before you teach

- Are children ready to approach the abstract concepts in this lesson?
- What will you provide for those children still needing concrete materials or pictorial representations?

NATIONAL CURRICULUM LINKS

Year 1 Number – number and place value

Identify and represent numbers using objects and pictorial representations including the number line, and use the language of: equal to, more than, less than (fewer), most, least.

ASSESSING MASTERY

Children can recognise numbers and link them to the amount they represent. Children can compare and order numbers represented in more abstract ways, using representations and resources they think most appropriate.

COMMON MISCONCEPTIONS

Children may confuse digits and the amount they represent. Ask:
- *What tools have you used in the past to help you with what a number is worth?*

Children may confuse > (greater than) and < (less than). Ask:
- *Which side do you start reading from?*
- *Does the sign begin with the bigger end or the smaller end?*

STRENGTHENING UNDERSTANDING

Provide number cards, each of which shows an amount represented using a number track or towers of multilink cubes, as well as the large >, < and = signs used in Lesson 11.

GOING DEEPER

Give children two ten-sided dice each. Ask children to roll the dice and order the numbers that they have rolled using the <, > and = signs.

KEY LANGUAGE

Other language to be used by the teacher: <, >, =, represent, more than, fewer than, greater than, how many, same, compare

STRUCTURES AND REPRESENTATIONS

Number track

RESOURCES

Mandatory: multilink cubes, digit cards

Optional: ten-sided dice, ten frames, counters, bead strings, large printed versions of the <, > and = signs for children to hold and manipulate, a selection of countable real-life objects such as pencils, a classroom display of the language and comparison signs

 In the eTextbook of this lesson, you will find interactive links to a selection of teaching tools.

Quick recap

Give children (in pairs or groups) the numbers from 1 to 10 on cards. Mix them up and ask them to put the numbers in order.

Discover

WAYS OF WORKING Pair work

ASK

- Question ❶ a): *Where is Bo's number on the track? Where is Jess's number on the track?*
- Question ❶ b): *Which number is greater? How do you know?*

IN FOCUS Question ❶ b) helps children to start comparing numbers using only numerals. The use of a number track can support children as they should know that the numbers increase in size as they move over to the right, so the position of numbers can help their comparisons.

PRACTICAL TIPS Encourage children to count the numbers on the number track from left to right to start with. Are they counting forwards or backwards? Are the numbers getting bigger or smaller? This can then help them make sense of their answers.

ANSWERS

Question ❶ b): Jess's number, 7, is greater than Bo's number, 4.

Compare numbers

Discover

❶ a) Point to Bo's number.
Point to Jess's number.

b) Which number is greater?

52

PUPIL TEXTBOOK 1A PAGE 52

Share

WAYS OF WORKING Whole class teacher led

ASK

- Question ❶ a): *How do the towers of cubes help you?*
- Question ❶ a): *How does the number track help?*
- Question ❶ b): *How do you know 7 is greater than 4?*
- Question ❶ b): *How do you know what sign to use?*

IN FOCUS Question ❶ helps children to begin to move from comparing concrete amounts to the abstract comparison of numerals. The cubes link back to previous learning, whilst the transition to the number track will support them in working more abstractly.

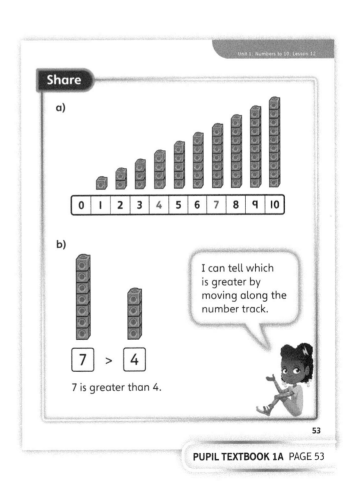

53

PUPIL TEXTBOOK 1A PAGE 53

Think together

WAYS OF WORKING Whole class teacher led (I do, We do, You do)

ASK

- Question **1**: *What will you need to do first to compare the amounts?*
- Question **2**: *What sign will you need to use? How can you prove that the sign you have chosen is the correct one? Can you prove it in another way?*
- Question **2**: *How can you show your comparison of the scores?*

IN FOCUS Question **2** asks children to compare two scores. Be mindful of the potential misconceptions regarding the use of the >, < and = signs and also recognising the value of the digits. Reinforce children's use of different representations and the tools that they have used in previous lessons.

STRENGTHEN Have a display in the classroom that shows the language and comparison signs, so that children can refer to the display as a reminder.

DEEPEN In question **3**, use Flo's comment about using multilink cubes to help her as an opportunity to discuss any other resources or representations that children might use. Using a variety of representations helps children to develop mathematical fluency and confidence.

ASSESSMENT CHECKPOINT Questions **2** and **3** will help you to decide if children can relate a digit to the amount it represents. Can children compare the amounts in the questions and explain the comparison using the correct language and signs?

ANSWERS

Question **1**: 5 < 8

Question **2**: 5 > 3 or 3 < 5

Question **3** a): 8, 9 and 10 are greater than 7.

Question **3** b): 0, 1, 2, 3, 4, 5 and 6 are less than 7.

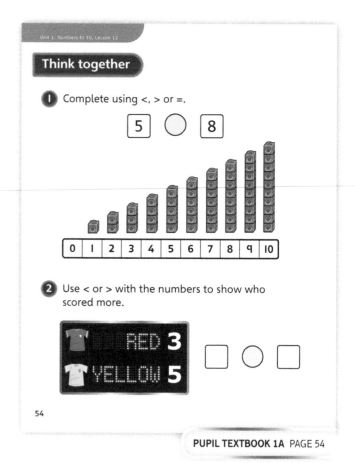

PUPIL TEXTBOOK 1A PAGE 54

PUPIL TEXTBOOK 1A PAGE 55

Practice

WAYS OF WORKING Independent thinking

IN FOCUS Questions **1**, **2** and **4** require children to compare only abstract digits and to demonstrate their understanding of the related concrete amounts. Question **3** helps children to practise comparing concrete amounts with abstract digits.

STRENGTHEN Ensure that children have access to all number resources and representations from the previous lessons to support their comparisons. Link the questions to contexts that children know and understand by asking questions, such as: *Would you rather have 6 sweets or 10 sweets? Why?*

DEEPEN In question **5** encourage children to think about Ash's comment and to work systematically to find all the solutions. Deepen their understanding by asking: *How many solutions can you find for* ▢ *> 6?*

ASSESSMENT CHECKPOINT Questions **1** and **2** should help you to decide if children recognise and understand digits. Questions **3** and **4** should help you to assess if children can use the >, < and = signs confidently and accurately.

ANSWERS Answers for the **Practice** part of the lesson appear can be found in the *Power Maths* online subscription.

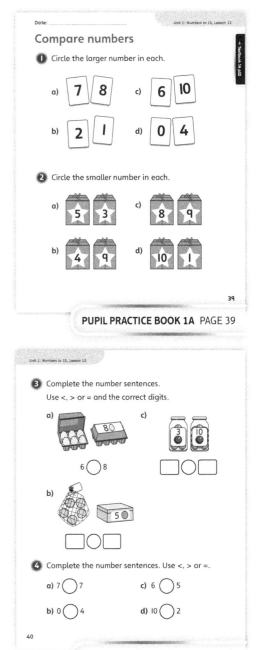

PUPIL PRACTICE BOOK 1A PAGE 39

PUPIL PRACTICE BOOK 1A PAGE 40

Reflect

WAYS OF WORKING Whole class

IN FOCUS Develop this **Reflect** activity using different patterns. For example, you could use the following repeated pattern: 'bigger, bigger, smaller, bigger, bigger, smaller'. If children select 10 as the first 'bigger' in the repeated pattern, use this as an opportunity to begin discussing what is bigger than 10.

ASSESSMENT CHECKPOINT This **Reflect** activity should help you assess children's understanding of greater than and smaller than, as well as their understanding and fluency with numbers to 10.

ANSWERS Answers for the **Reflect** part of the lesson can be found in the *Power Maths* online subscription.

After the lesson

- Did moving from concrete materials and pictorial representations to the abstract concepts in this lesson result in any unexpected misconceptions?
- How will you approach these misconceptions in Lesson 13?

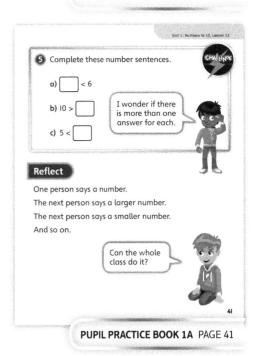

PUPIL PRACTICE BOOK 1A PAGE 41

93

Order objects and numbers

Learning focus

In this lesson, children will compare three or more groups of objects and order them in both ascending and descending order.

Before you teach

- What challenges will you set children who can already compare more than two groups?
- Is there a representation of amounts that children are relying on more than other representations? How could you increase their confidence with the other representations?

NATIONAL CURRICULUM LINKS

Year 1 Number – number and place value

Identify and represent numbers using objects and pictorial representations including the number line, and use the language of: equal to, more than, less than (fewer), most, least.

ASSESSING MASTERY

Children can confidently count groups of objects and recognise the comparative place value of numbers to 10. Children can compare more than two amounts and can arrange these amounts in both ascending and descending order, justifying their ideas using pictorial representations and concrete materials.

COMMON MISCONCEPTIONS

Children may arrange numbers in ascending order when asked to arrange them in descending order. Ask:
- What does 'most' mean?
- Which number is the greatest?

STRENGTHENING UNDERSTANDING

Provide number cards that show pictured examples of the concrete materials used in the lesson. For example, a card could show the digit, the appropriate face of a dice, the number represented on a ten frame, and so on. Link these number cards to a large number track displayed in the classroom. Consider children building the numbers using cubes and then asking them to stand the cubes in height order. This will help them put the numbers in ascending or descending order.

GOING DEEPER

When children are confident ordering numbers from smallest to largest or vice versa, ask them to suggest numbers that can go between two other numbers e.g., 2, ☐, ☐, 8. Ask: *What could the missing numbers be? Could they be the same?*

KEY LANGUAGE

In lesson: most, fewest, smallest, greatest
Other language to be used by the teacher: compare, order, least

STRUCTURES AND REPRESENTATIONS

Ten frame

RESOURCES

Mandatory: multilink cubes, ten frames, six-sided dice

Optional: number cards that show pictured examples of the concrete materials used in the lesson, a large number track on display in the classroom

 In the eTextbook of this lesson, you will find interactive links to a selection of teaching tools.

Quick recap

Ask children to roll a dice or pick a number from a hat. Ask them to compare their number with their partner's number. Who has the largest number? Who has the smallest?

Discover

Order objects and numbers

Discover

Kat   Em   Josh

WAYS OF WORKING Pair work

ASK

· Question ⓵: *Who do you predict has more stickers and why?*
· Question ⓵ a): *How many more stickers does Kat need before she has the same amount as Josh?*
· Question ⓵ b): *How many fewer stickers does Josh have in comparison to Em?*
· Question ⓵ b): *Can you order the amounts of stickers from greatest to least?*

IN FOCUS Questions ⓵ a) and ⓵ b) introduce the lesson's key language: 'most' and 'fewest'. Discuss ordering amounts from fewest to most and from most to fewest.

PRACTICAL TIPS Use pictures of stars at the front of the class to recreate the **Discover** scenario. Alternatively, give children pictures of stars and ask them to recreate the problem in small groups.

ANSWERS

Question ⓵ a): Em has the most stars.

Question ⓵ b):

4	6	7
Kat	Josh	Em

❶ a) Who has the most stars?

b) Put the children's stars in order from fewest to most.

56

PUPIL TEXTBOOK 1A PAGE 56

Share

WAYS OF WORKING Whole class teacher led

ASK

· Question ❶: *What should you do before you compare the amounts?*
· Question ❶ a): *Why has Astrid chosen to use multilink cubes?*
· Question ❶ a): *What will you use to represent the numbers? Explain why.*
· Question ❶ b): *What can you tell me about the numbers?*
· Question ❶ b): *Can you make a number sentence using the >, < and = signs you have learnt about?*

IN FOCUS Question ❶ a) and ❶ b) encourage children to order amounts from least to greatest when comparing amounts. Refer to Flo's question to assess if children can explain what 'most' means. Encourage children to make the numbers using towers of cubes to compare heights so enabling them to order numbers. The smallest tower would be the smallest height and so on.

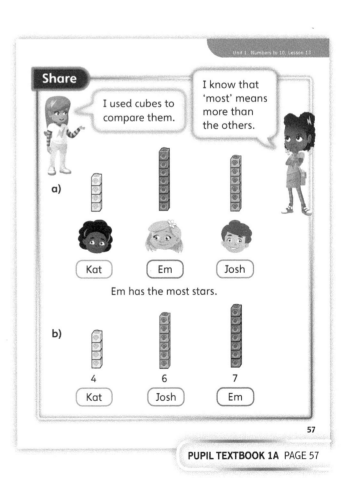

PUPIL TEXTBOOK 1A PAGE 57

Think together

WAYS OF WORKING Whole class teacher led (I do, We do, You do)

ASK

- Question **1**: *Which tower is the shortest? Which is the smallest number? Which tower is the tallest? Which is the greatest number?*
- Question **2**: *What could you use to make the numbers?*

IN FOCUS Question **1** introduces children to the key language of 'smallest' and 'greatest'. It also develops their understanding of ordering numbers as they use towers of cubes to write numbers in order. In question **2**, children could try to order the numbers using their knowledge of numbers on the number track or by making towers of cubes to reinforce their understanding.

STRENGTHEN Children should be given access to different colours of multilink cubes, so they can build each number to support them. Ask: *How does the height of the tower help you order the number? If you are starting with the smallest number, which tower comes first? How do you know?*

DEEPEN In question **3**, children are required to order numbers that are presented in different ways. They should be encouraged to explain their methodology. Ask: *How do you know where Eve's and Toby's numbers are on the number tracks? What did you do first? What do you notice about the position of the numbers on the number track? How does that help you to order the numbers?*

ASSESSMENT CHECKPOINT Questions **1** and **2** should help you to assess whether children can write numbers in order from smallest to greatest or greatest to smallest.

ANSWERS

Question **1** a): Adam has the smallest number.

Question **1** b): Bob has the greatest number.

Question **2**: 1, 7, 10

Question **3** a): Children point to 4, 3 and 5 on the number track.

Question **3** b): Toby's score of 5 is the greatest.

Question **3** c): Eve's score of 3 is the least.

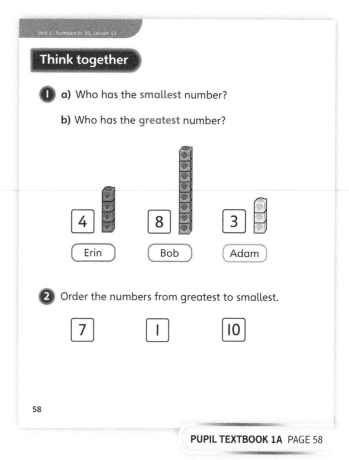

Think together

1 a) Who has the **smallest** number?

b) Who has the **greatest** number?

4	8	3
Erin	Bob	Adam

2 Order the numbers from greatest to smallest.

| 7 | 1 | 10 |

58

PUPIL TEXTBOOK 1A PAGE 58

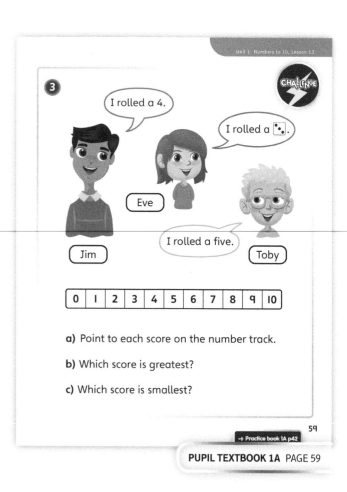

3

I rolled a 4.

I rolled a ⚁.

Eve

Jim

I rolled a five.

Toby

| 0 | 1 | 2 | 3 | 4 | 5 | 6 | 7 | 8 | 9 | 10 |

a) Point to each score on the number track.

b) Which score is greatest?

c) Which score is smallest?

59

→ Practice book 1A p42

PUPIL TEXTBOOK 1A PAGE 59

Practice

WAYS OF WORKING Independent thinking

IN FOCUS Questions **1** and **2** allow children to demonstrate their understanding of the word 'smallest'. The questions move from pictorial representations of objects to the abstract numerals. Questions **3** and **4** follow the same structure to enable children to demonstrate their understanding of greatest.

STRENGTHEN Provide children with different colours of multilink cubes, so that they can represent the numbers in the questions to support their thinking.

DEEPEN In question **5**, encourage children to compare their answers with a partner. Ask: *Have you ordered them the same way? Could you order them another way? What is the same and what is different? With parts a) to c), what do you notice about the middle number whichever way you ordered the numbers?*

ASSESSMENT CHECKPOINT Questions **1** to **4** help you to assess whether or not children can interpret the key language of greatest and smallest, both pictorially and abstractly, whilst in question **5** children demonstrate whether they are able to write numbers in order.

ANSWERS Answers for the **Practice** part of the lesson can be found in the *Power Maths* online subscription.

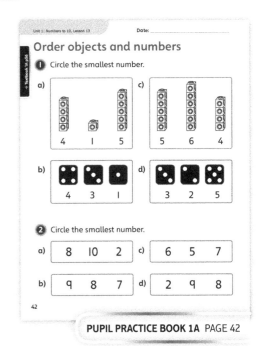

PUPIL PRACTICE BOOK 1A PAGE 42

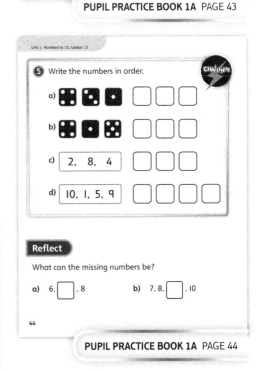

PUPIL PRACTICE BOOK 1A PAGE 43

Reflect

WAYS OF WORKING Independent thinking

IN FOCUS This activity provides the opportunity for children to think about how the numbers have been ordered to decide what the missing number is.

ASSESSMENT CHECKPOINT Children should be able to identify the missing number in each set and explain how they came to their answer.

ANSWERS Answers for the **Reflect** part of the lesson can be found in the *Power Maths* online subscription.

After the lesson ⏸

- Did increasing the number of groups being compared create any unexpected issues? If so, what were the issues and how did they change your lesson?
- Are children able to arrange numbers in both ascending and descending order?

PUPIL PRACTICE BOOK 1A PAGE 44

The number line

Learning focus

In this lesson, children will learn to recognise and use the representation of a number line to help them answer questions based on all of the learning in this unit.

Before you teach

- How will you make the links between this representation and the previous representations clear?
- Could these links be displayed in the classroom to help provide scaffolding for children's understanding?

NATIONAL CURRICULUM LINKS

Year 1 Number – number and place value

Identify and represent numbers using objects and pictorial representations including the number line, and use the language of: equal to, more than, less than (fewer), most, least.

ASSESSING MASTERY

Children can use a number line to help them answer questions based on the learning that has taken place throughout Unit 1. Children can explain how a number line works, how it represents numbers and amounts, and how it helps them to compare, order and count up and down from different numbers.

COMMON MISCONCEPTIONS

Children may not recognise that a number line features 0 as an amount. This is different to the number track and the ten frame that children have met so far, because these representations only start recording amounts at 1. Ask:
- *What number does the number track start at?*
- *What number does the number line start at?*
- *How are they different?*

Give children groups that contain different amounts of objects and ask children to point to the correct number on the number line. Give them a group of 0 things and ask:
- *Where would you place this group on a number line?*
- *Could you represent this group on the number line if the 0 wasn't on the number line?*

STRENGTHENING UNDERSTANDING

To introduce the concept of the number line, play a simple dice game along a number line. Start at 0 and roll a six-sided dice to move up the number line. This game could be played outside with a large number line drawn in chalk on the ground.

GOING DEEPER

Give children a template to create their own missing number questions for a partner to complete using a number line. Ensure that children check their own questions before giving them to their partner.

KEY LANGUAGE

In lesson: number line

Other language to be used by the teacher: order, greater, one more, one less

STRUCTURES AND REPRESENTATIONS

Number line

RESOURCES

Mandatory: number line

Optional: six-sided dice, chalk, multilink cubes, ten frames, number tracks, a selection of countable real-life objects such as bags or pencil cases flashcards of numbers 1 to 10, string, pegs

 In the eTextbook of this lesson, you will find interactive links to a selection of teaching tools.

Quick recap 🔾

Play a counting game or sing a song to ensure children can count forwards and backwards from 0 to 10.

Discover

Unit 1: Numbers to 10, Lesson 14

The number line

Discover

WAYS OF WORKING Pair work

ASK
- Question ❶ a): *What numbers do you recognise?*
- Question ❶ b): *Do the numbers match how many bags there are?*

IN FOCUS Questions ❶ a) and ❶ b) offer children their first opportunity to work with numbers on a number line. Question ❶ a) recaps their understanding of cardinal numbers by requiring them to recognise the missing numbers.

PRACTICAL TIPS Hold a string along the front of the class with pegs holding up numbers set out in the same way as the **Discover** artwork.

ANSWERS

Question ❶ a): 4, 8 and 9 are missing.

Question ❶ b): 3 has fallen down. 6 and 7 are in the wrong order.

❶ a) What numbers are missing?

b) What else is wrong?

PUPIL TEXTBOOK 1A PAGE 60

Share

WAYS OF WORKING Whole class teacher led

ASK
- Question ❶ a): *How did you know what numbers were missing?*
- Question ❶ b): *How did you know that 6 and 7 were in the wrong order?*
- Question ❶: *Why is this called a number line?*
- Question ❶: *How is this number line similar to the representations you have used before? What is different about it?*
- Question ❶: *Can you see any patterns in the number line?*

IN FOCUS Flo introduces the lesson's key language: 'number line'. Discuss the fact that the number line begins at 0.

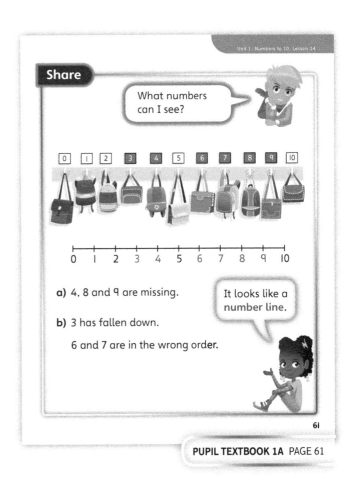

Share

What numbers can I see?

a) 4, 8 and 9 are missing.

It looks like a number line.

b) 3 has fallen down.

6 and 7 are in the wrong order.

PUPIL TEXTBOOK 1A PAGE 61

Think together

WAYS OF WORKING Whole class teacher led (I do, We do, You do)

ASK

• Question **1**: *What number comes next? How do you know?*
• Question **2**: *What can you do first to help you?*

IN FOCUS Question **1** reinforces children's understanding of how a number line is constructed and what patterns can be found. Children should count aloud when labelling the number line. In question **2**, they should recognise that they need to count along 6 intervals and that labelling the number line first will help them.

STRENGTHEN Children could build towers of multilink cubes above the number line in a similar way to how they used number tracks to support this transition.

DEEPEN In question **3**, children use a number line to identify one more and one less. Encourage children to explain how Astrid's jumps on the number line help her to find the answer. Ask: *What is the same and what is different about finding one more and finding one less? Can you use the number line to complete the task set by Sparks?*

ASSESSMENT CHECKPOINT Questions **1** and **2** will help you to assess whether children can accurately label a number line. Ensure they write the numbers in the correct position and not within the intervals.

ANSWERS

Question **1**: 6, 7, 8, 9, 10

Question **3** a): One more than 6 is 7.

Question **3** b): One less than 6 is 5.

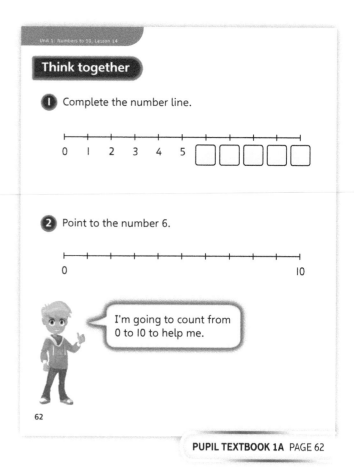

PUPIL TEXTBOOK 1A PAGE 62

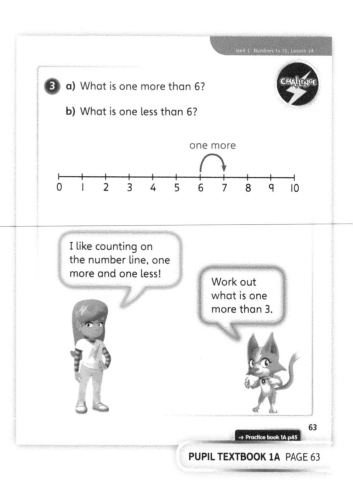

PUPIL TEXTBOOK 1A PAGE 63

Practice

WAYS OF WORKING Independent thinking

IN FOCUS Questions ① and ② allow children to demonstrate their understanding of the number line and how numbers are arranged along it. The questions provide decreasing levels of scaffolding and require children to visualise more of the numbers on the unlabelled divisions. Questions ③ and ④ allow children to use the number line to practise skills learnt in previous lessons.

STRENGTHEN If children complete the number line in question ① incorrectly, show them a printed example of a correct number line and ask them to identify what is the same and what is different between their line and the printed line.

DEEPEN In question ⑤ use Flo's comment to discuss how using the number line can be a useful thing to do. Ask: *How might it help you?*

ASSESSMENT CHECKPOINT Questions ③ to ⑤ should help you assess whether children can use the number line fluently to compare the size of numbers to 10 and to count forwards and backwards by 1 or more.

ANSWERS Answers for the **Practice** part of the lesson can be found in the *Power Maths* online subscription.

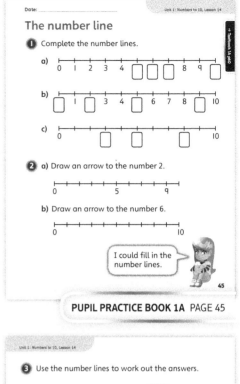

PUPIL PRACTICE BOOK 1A PAGE 45

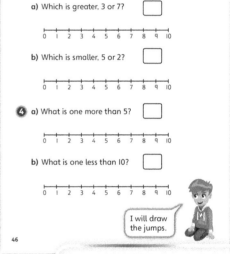

PUPIL PRACTICE BOOK 1A PAGE 46

Reflect

WAYS OF WORKING Independent thinking

IN FOCUS This **Reflect** activity allows children to reflect on what they have learnt in this lesson and link it to concepts that they have learnt throughout Unit 1. Discuss what children feel they have learnt to use the number line to do. Point out the different uses that children mention.

ASSESSMENT CHECKPOINT Assess whether children can identify uses for the number line and confidently use the appropriate language to explain their ideas.

ANSWERS Answers for the **Reflect** part of the lesson can be found in the *Power Maths* online subscription.

After the lesson

- Are children able to confidently apply their knowledge and understanding to the new representation? Did this lesson highlight any lingering misconceptions around the concepts taught in previous lessons?
- Did children have enough opportunities to explain their reasoning about number lines?

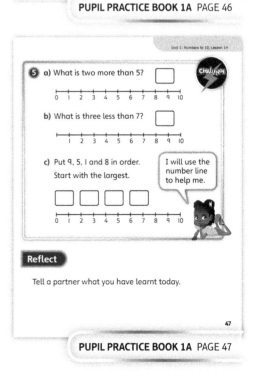

PUPIL PRACTICE BOOK 1A PAGE 47

End of unit check

> Don't forget the unit assessment grid in your *Power Maths* online subscription.

WAYS OF WORKING Group work adult led

IN FOCUS

- Questions **1**, **3** and **5** focus on children's understanding of counting forwards and backwards from numbers up to 10.
- Question **2** assesses children's understanding of the ten-frame representation of numbers up to 10.
- Question **4** assesses children's ability to compare two numbers and describe their relationship to each other.

Think!

WAYS OF WORKING Pair work or small groups

IN FOCUS This question assesses children's ability to recognise and compare numbers to 10. They should be able to recognise that 6 has been partitioned in two different ways, giving three amounts to compare (the red balloons, the yellow balloons and all the balloons). Ask:

- *What is the same and different about the two sets of balloons?*
- *What numbers can you see in the pictures?*
- *Who has more red balloons? How do you know?*

Draw children's attention to the words at the bottom of the **My journal** page and encourage them to use them in their answers.

Encourage children to think through or discuss how many balloons Bea and Seth have, and what they can say about those balloons, before writing their answer in **My journal**.

ANSWERS AND COMMENTARY Children will demonstrate mastery in this concept by recognising and comparing the different numbers they can see in the picture. They will count reliably and fluently and use language such as 'more', 'less' and 'equal'. If asked to, they should be able to choose an appropriate representation for the balloons and use this to help prove their comparisons.

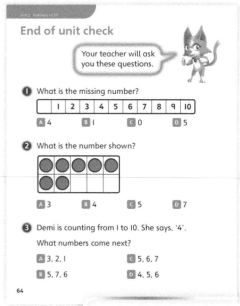

PUPIL TEXTBOOK 1A PAGE 64

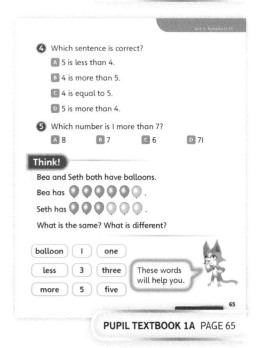

PUPIL TEXTBOOK 1A PAGE 65

Q	A	WRONG ANSWERS AND MISCONCEPTIONS	STRENGTHENING UNDERSTANDING
1	C	Choosing B suggests that the child is not confident with the concept of 0.	To help children gain fluency in counting on and back, give them a number track or number line and ask: • *Can you point to the number you are starting at?* • *Which way will you move along the track/line if you are counting forward/back?* Give children opportunities to compare numbers using concrete resources. Ask: • *How could you represent the numbers?* • *What do you notice about them?* • *Which number is greater? How do you know?*
2	D	Choosing A suggests that the child has counted the blank spaces.	
3	C	Choosing D indicates that the child has recognised they need to count forward but has started at 4, not the number following it.	
4	D	Choosing A or B could indicate a lack of understanding of the value of the numbers or the vocabulary of 'more' and 'less' (also Q5 B, C).	
5	A	Choosing C suggests that the child has counted one less, rather than one more.	

My journal

WAYS OF WORKING Independent thinking

ANSWERS AND COMMENTARY

What is the same?
- Each child has 6 balloons.
- Both children have spotted and plain balloons.

What is different?
- Bea has 2 more spotted balloons than Seth.
- Seth has 2 fewer spotted balloons than Bea.
- Bea has 5 spotted balloons and Seth has 3.

- Seth has 2 more plain balloons than Bea.
- Bea has 2 fewer plain balloons than Seth.
- Bea has 1 plain balloon and Seth has 3.

If children are finding it difficult to articulate anything but surface similarities and differences (for example, 'They are different patterns'), ask:
- *Do they have the same number of spotted/plain balloons? How do you know?*
- *Who has more spotted/plain balloons? Can you prove it?*

Power check

WAYS OF WORKING Independent thinking

ASK
- *How confident do you feel when counting up/back to/from 10?*
- *Do you prefer using a number track or a number line? Why?*

Power play

WAYS OF WORKING Pair work or small groups

IN FOCUS Use this **Power play** to assess whether children are fluent in making numbers up to 10 using the ten frame. Children should be able to recognise and demonstrate that any number below 10 can be arranged in different ways and still be worth the same amount.

ANSWERS AND COMMENTARY Children should realise that 9 is the biggest number they can make. They will show through their discussion that they can compare two numbers, explaining how they know which is bigger and which is smaller. They will recognise that a number can be presented in different ways on a ten frame but still keep its same value.

PUPIL PRACTICE BOOK 1A PAGE 48

PUPIL PRACTICE BOOK 1A PAGE 49

After the unit ⏸

- In what other areas of the curriculum could you use the skills learnt in this unit? For example: comparisons of measurements in science, counting down to start a race in PE.
- Were children honest in their assessment of their own ability? How could you develop their ability to self-assess in future lessons?

Strengthen and **Deepen** activities for this unit can be found in the *Power Maths* online subscription.

Unit 2
Part-whole within 10

Don't forget to watch the Unit 2 video!

Mastery Expert tip! 'When I taught this unit I found it was worth spending a good amount of time ensuring children could confidently identify the parts and the whole before starting to deal with finding missing numbers. Taking my time with the early lessons meant my children ended up making more rapid progress in the later lessons.'

WHY THIS UNIT IS IMPORTANT

This is one of the most important units of work to teach in maths. Children who have a solid grasp of numbers to 10 are able to apply this knowledge to so many other areas, including numbers to 20, numbers to 50 and beyond, and addition and subtraction.

Conversely, children who are not secure with the idea of partitioning within 10 may end up constantly playing catch up, due to the vast number of topics that rely on this knowledge. It is therefore worth spending time exploring this concept in depth.

WHERE THIS UNIT FITS

→ Unit 1: Numbers to 10
→ **Unit 2: Part-whole within 10**
→ Unit 3: Addition within 10

This unit, which builds on Unit 1: Numbers to 10, introduces children to the part-whole model, focusing on different ways of partitioning numbers to 10. Children use the part-whole model to help them write and compare number bonds. They will continue to use these skills in Unit 3, which focuses on addition.

Before they start this unit, it is expected that children:
- know how to sort and compare objects to 10
- understand how to count on and back within 10
- can order a set of numbers and use the vocabulary 'less than', 'more than' and 'equal to'.

ASSESSING MASTERY

Children who have mastered this unit will be able to confidently partition numbers within 10 using a part-whole model. They can write the associated number sentences and be flexible with where they write the = sign; for example, they know that 3 + 2 = 5 and 5 = 3 + 2 represent the same fact.

COMMON MISCONCEPTIONS	STRENGTHENING UNDERSTANDING	GOING DEEPER
Children may confuse the parts and the whole.	Allow children to spend time recognising and identifying the parts and the whole in a part-whole model without any numbers in it. This means children only have one thing to focus on and learn.	Children could explore part-whole models with more than two parts. It is important to ensure that the whole remains within 10.
Children may write incorrect number sentences based on confused part-whole thinking; for example, thinking that 4 + 6 = 2 is the same as 4 + 2 = 6.	Encourage children to use cubes to prove their answers and reinforce the meaning of the = sign.	Ask children more open-ended questions, such as: *The whole is 8, what could the parts be? How many different number sentences can you write?*

UNIT STARTER PAGES

Introduce the unit using teacher-led discussion. Give children time to discuss each question in small groups or pairs and then discuss their ideas as a class.

STRUCTURES AND REPRESENTATIONS

Part-whole model: This model helps children understand that two or more parts combine to make a whole. It also helps to strengthen children's understanding of number.

Five frame and ten frame: The five and ten frames help to give children a sense of the numbers, and support their understanding of number bonds to 5 and 10. They also play a key role in helping children to recognise the structure of other numbers, and to understand what happens when you add two numbers together.

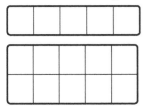

Bead string: The bead string is a great way of introducing children to patterns and helping them to be systematic in their approach. They can find answers by moving one bead at a time, each time recording the number sentence they have represented.

KEY LANGUAGE

There is some key language that children will need to know as part of the learning in this unit.

➔ part-whole model, part, whole, groups

➔ number sentence, number bonds

➔ plus

➔ is equal to

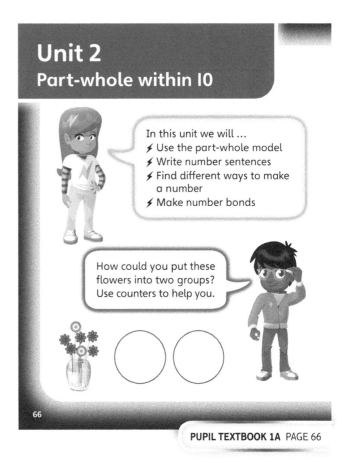

PUPIL TEXTBOOK 1A PAGE 66

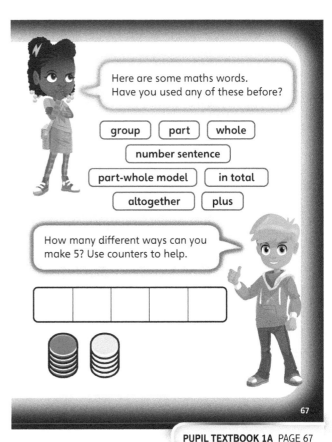

PUPIL TEXTBOOK 1A PAGE 67

Parts and wholes

Learning focus

In this lesson, children are starting to see that a whole group can be made up of two (or more) parts. Children will build on their knowledge of counting or subitising objects in a group.

Before you teach

- Think about the objects you are going to give children and how they may split them into groups.
- Look carefully at the examples in the **Think together** section for ideas of different parts and wholes. This may help if you need some other examples in the lesson.

NATIONAL CURRICULUM LINKS

Number – addition and subtraction

Identify and represent numbers using objects and pictorial representations including the number line, and use the language of: equal to, more than, less than (fewer), most, least.

Represent and use number bonds and related subtraction facts within 20.

ASSESSING MASTERY

Children can recognise a whole and parts that make up a whole. They should also be able to work out how many are in each group by counting or subitising and, therefore, work out how many are in the whole.

COMMON MISCONCEPTIONS

Children can sometimes struggle to see two groups when objects are close together or they are not distinctly different in some way. If this is the case, consider making sure that there is a gap between the two groups. Ask:
- *How many groups are there?*
- *How can you tell where one group ends and the other group begins?*

STRENGTHENING UNDERSTANDING

Use a group of real-life objects and ask children to put them together in a whole and ask them just to practise splitting the whole into two groups. Listen out for children saying things like 'This is the whole' and 'these are the parts'.

GOING DEEPER

Ways to go deeper include giving children six completely different objects and asking them to split them into two parts. This is more challenging because the objects are all different. You may also want to ask them to split a whole into three parts.

KEY LANGUAGE

In lesson: in total, part, whole

RESOURCES

Mandatory: objects from the classroom, cubes, counters

 In the eTextbook of this lesson, you will find interactive links to a selection of teaching tools.

Quick recap 🔁

Put ten sticks on the floor. Ask children to take one, two or three sticks. Keep playing the game until there are no more sticks to pick up.

Discover

Parts and wholes

Discover

WAYS OF WORKING Pair work

ASK

- Question ① a): *Can you see the two groups of frogs?*
- Question ① a): *Can you see how many frogs there are in each group?*
- Question ① b): *Did you have to count the frogs or could you work out how many frogs are in each group in other ways?*
- Question ① b): *Can you see the whole?*

IN FOCUS The purpose of this activity is that children start to recognise what the whole is and that it can be divided into two parts. Ask children to circle the whole with their finger and then the parts. Make sure you use the language parts and wholes.

PRACTICAL TIPS Use real-life objects to represent the frogs.

ANSWERS

Question ① a): There are 3 frogs on the log. There are 2 frogs in the water.

Question ① b): There are 5 frogs altogether.

① a) Can you see two groups of frogs?

How many frogs are in each group?

b) How many frogs are there in total?

68

PUPIL TEXTBOOK 1A PAGE 68

Share

WAYS OF WORKING Whole class teacher led

ASK

- Question ① a): *Can you see the three frogs and two frogs? What do you think the parts are?*
- Question ① b): *Can you see the five frogs in total? Can you see the whole?*

IN FOCUS In question ① a), a key focus is that the parts are in two circles. This will help them when they meet the part-whole model in the next lesson. Sparks's comment is a key sentence stem that children should use throughout this lesson. Ask children to say 'Three is a part, two is a part. The whole is five.' as a group. In order to find the parts and whole we are asking children to count or subitise. Note that this lesson is not about adding two numbers together, this is done in Unit 3. This is about parts and wholes.

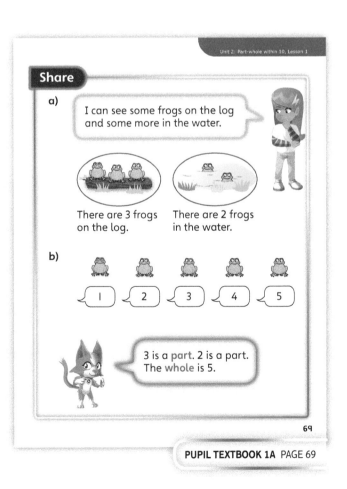

PUPIL TEXTBOOK 1A PAGE 69

Think together

WAYS OF WORKING Whole class teacher led (I do, We do, You do)

WAYS OF WORKING Whole class teacher led (I do, We do, You do)

ASK

- Question **1**: *How many cookies are on each plate? What are the parts?*
- Question **2**: *Where do you think the parts are in this example?*
- Question **3**: *What are the objects? Can you put the objects into two groups? How many ways can you do this?*

IN FOCUS In this section, it is important for children to see the two parts that make up the whole. Ensure children can show you this by circling with their fingers. Reinforce the use of the stem sentences throughout of 'X is a part, Y is a part and Z is the whole'. Questions **1** and **2** are the key questions that children need to ensure they understand. You may want to add some examples where children draw their own groups of objects and work out how many there are in each part and in the whole before they move on to the challenge question.

STRENGTHEN Make sure children circle the parts with their fingers.

DEEPEN If children are counting objects, try and encourage them to subitise. Question **3** is a great example where children have four different objects and they have to split them into two groups in different ways. They may struggle to know that there is more than one way to do this. Each time they should use the stem sentence. Ask them what they notice each time.

ASSESSMENT CHECKPOINT Make sure children can look at a whole group of objects and see two groups. They should be confident using the stem sentence 'X is a part, Y is a part and Z is the whole'. Ask them to say this frequently.

ANSWERS

Question **1**: There are 2 cookies on one plate and 4 cookies on the other plate.

Question **2**: There are 3 bugs in the one part and 1 bug in the other part.

Question **3** a): The cubes could be split as 2 + 2 or 3 + 1 or 1 + 3 or 4 + 0 or 0 + 4.

Question **3** b): Answers will depend on part a). For example: 3 is a part and 1 is a part. The whole is 4 for every combination.

Think together

1 How many cookies on each plate?

2 How many bugs in each part?

I wonder if I can work out the wholes.

70

PUPIL TEXTBOOK 1A PAGE 70

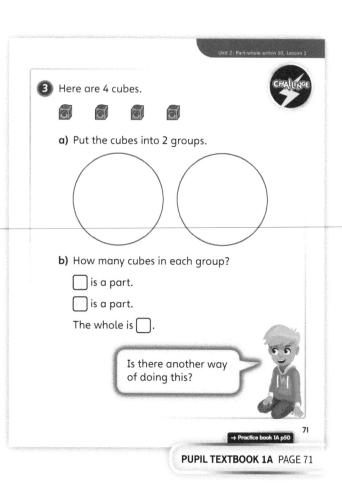

3 Here are 4 cubes.

CHALLENGE

a) Put the cubes into 2 groups.

b) How many cubes in each group?

☐ is a part.

☐ is a part.

The whole is ☐.

Is there another way of doing this?

71

→ Practice book 1A p50

PUPIL TEXTBOOK 1A PAGE 71

Practice

IN FOCUS

Question ① focuses on children just counting or subitising how many objects there are in each part. They see the circles around each part. Questions ② and ③ include the whole and ask children to complete the stem sentences. They should now be getting confident with this stem sentence. Question ⑤ is the first time children break up something into two parts and they can see clearly that an object can be broken into parts in different ways.

STRENGTHEN Use cubes and counters for children to represent some of the objects and put them on paper plates or circles. Ask them to count the objects and say the stem sentence. Point to each of the parts as they say the sentence out loud.

DEEPEN Build on question ⑤ by using more cubes and asking children to break towers of cubes into different parts. Ask them to record their answers.

ASSESSMENT CHECKPOINT Can children recognise and count how many objects there are in each group? Can they use the stem sentence correctly?

ANSWERS Answers for the **Practice** part of the lesson can be found in the *Power Maths* online subscription.

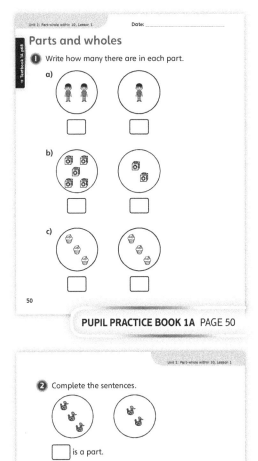

PUPIL PRACTICE BOOK 1A PAGE 50

PUPIL PRACTICE BOOK 1A PAGE 51

Reflect

WAYS OF WORKING Independent thinking

IN FOCUS This task allows children to draw their own two parts. They may choose to make the parts from objects in the classroom or on their tables, such as cubes and counters. Ask children to use the stem sentence for their parts and wholes.

ASSESSMENT CHECKPOINT By the end of the lesson, children should be able to confidently use the stem sentence provided. Ask them to use it as a whole class on some of the part-wholes they have created.

ANSWERS Answers for the **Reflect** part of the lesson can be found in the *Power Maths* online subscription.

After the lesson ⏸

- Are children able to use the stem sentence?
- Can children draw two parts for a whole?

PUPIL PRACTICE BOOK 1A PAGE 52

The part-whole model

Learning focus

In this lesson children learn that a number can be partitioned into two parts using a part-whole model. Children explore that numbers can be partitioned in different ways.

Before you teach

- Are all children secure with the idea of one-to-one correspondence used in numbers to 10?
- What resources will you provide for children who are still developing these ideas?

NATIONAL CURRICULUM LINKS

Number – addition and subtraction

Represent and use number bonds and related subtraction facts within 20.

ASSESSING MASTERY

Children can partition numbers to 10 using a part-whole model.

COMMON MISCONCEPTIONS

Children may get the numbers mixed up. For example, they may write 5 as the whole and 6 and 1 as the parts. Ask:
- *In these part-whole models, can you point to the whole? Can you point to the parts?*
- *Where should the biggest number go?*

STRENGTHENING UNDERSTANDING

Use hoops in the playground to introduce the idea of the part-whole model. Children can stand in the hoops and experiment with different combinations. If 7 is the whole, how many different ways can children stand in the hoops?

GOING DEEPER

To extend the practical activity above, children can decide what they should do if one child steps out of one of the hoops. Can the other children stay where they are? Does someone have to leave both of the other hoops or just one of them? What if two children leave each part? How many children need to leave the hoop representing the whole?

KEY LANGUAGE

In lesson: part-whole model

Other language to be used by the teacher: group, whole, part, diagram, different, partition, biggest, true, false

STRUCTURES AND REPRESENTATIONS

Part-whole model

RESOURCES

Mandatory: counters

Optional: hoops, teddy bears, selection of countable objects

 In the eTextbook of this lesson, you will find interactive links to a selection of teaching tools.

Quick recap

Give children five counters and five cubes. Ask them to show three counters and two cubes. Check they have done it. Make other numbers using counters and cubes.

Discover

The part-whole model

Discover

WAYS OF WORKING Pair work

ASK

- Question ① a): *Can you use any of the equipment on your table to show what the picture shows?*
- Question ① b): *Do you both agree what each number in the part-whole model shows?*
- Question ① b): *Can you explain to your partner how you got your answer?*

IN FOCUS In this part of the lesson children begin to learn about the idea that a number can be partitioned into two groups. They should think about and discuss with their partner what each number in the part-whole model shows.

PRACTICAL TIPS Recreate the scenario using children in hoops in the playground or school hall. Alternatively, you could ask children to work in small groups in the classroom and model the problem using teddy bears in hoops.

ANSWERS

Question ① a): There are 2 children in the red hoop and 4 children in the blue hoop.
There are 6 children altogether.

Question ① b): 6 is the whole and 2 and 4 are the parts.

① **a)** How many children are there in each group?

How many children are there in total?

b) What does each number in the part-whole model show?

72

PUPIL TEXTBOOK 1A PAGE 72

Share

WAYS OF WORKING Whole class teacher led

ASK

- *What did you use to show the part-whole model of the children?*
- *Which number represents the whole?*
- *Which numbers represent the parts?*
- *Would it still work if you put the 4 in the circle on the left and the 2 in the circle on the right? Is this still correct? (Point to each circle as you say this so children know which circles you are talking about.)*
- *Can I swap the 2 and 6 around? Is the part-whole model still correct?*

IN FOCUS Here children are looking at the part-whole model in more detail. They should be able to recognise which number is the whole and which numbers represent the parts. They also begin to experiment with which numbers can be moved and which numbers cannot.

PUPIL TEXTBOOK 1A PAGE 73

Think together

WAYS OF WORKING Whole class teacher led (I do, We do, You do)

ASK

- Question **1**: *What numbers go in the parts? What number does each part represent? What number goes in the whole? What does the whole represent?*
- Question **2**: *What do you notice about the parts?*

IN FOCUS Question **1** is about making sure that children can complete a part-whole model from the images provided. It is vital that children understand where the parts and the whole are. Be as clear as possible to ensure children know what each number represents in the part-whole model.

Question **2** provides children with a discussion opportunity around a part-whole model where the parts are the same.

STRENGTHEN Ask children to represent the children or objects using cubes or counters and draw large circles around the groups. Ask them to use the stem sentences from the previous lesson where they say the parts and the whole. This will help make it clear what each of the parts and the whole represent.

DEEPEN Question **3** allows children to start making their own groups. They can use a pencil to split the counters into two groups and draw a part-whole model that reflects this. Ask children to think if there is a systematic way of finding all the part-wholes for a possible whole. Get them to think about 0 as one of the parts. You could increase the number of groups.

ASSESSMENT CHECKPOINT By the end of this section, children should be able to draw a complete part-whole model for a group.

ANSWERS

Question **1**:

Question **2**:

Question **3**: The counters could be arranged as:
4 and 1 or 1 and 4
3 and 2 or 2 and 3
In this situation, we cannot use 0 and 5 as the counters would then still be in one group, but it would be correct for an abstract part-whole.

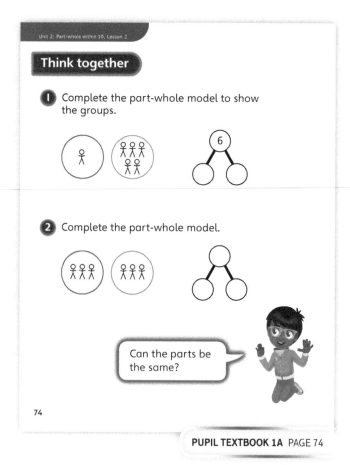

PUPIL TEXTBOOK 1A PAGE 74

PUPIL TEXTBOOK 1A PAGE 75

112

Practice

WAYS OF WORKING Independent thinking

IN FOCUS Children move between pictorial and abstract representations to show how numbers within 10 can be made up. They show their answers in the form of a part-whole model. They know that it shows the whole and parts.

STRENGTHEN Encourage children to use cubes or counters to make each part-whole model. The questions could be linked to familiar situations to help them understand groups. For example, at a teddy bears' picnic, three teddy bears sit in one group and four teddy bears sit in another group. How many teddy bears are there altogether? Use real teddy bears to show the groups.

Question ③ might also help to strengthen understanding. Children are given the whole in the part-whole model, and can check that the two parts, shown as counters in the circles, add up to the whole.

DEEPEN Give children a number and ask them to show you all the possible part-whole models they can for this number. Question ⑤ is an example of this. Ask: *Will each number make the same number of part-whole models? Which number (to 10) makes the greatest number of part-whole models? Is it possible to make a part-whole model for the number 1?*

ASSESSMENT CHECKPOINT Questions ① and ② check that children can read and interpret a part-whole model. Check that children are able to link a picture to an abstract model.

The numbers in question ④ have been chosen intentionally. The whole is the same in the first two models with one missing part, so children may be tempted to fill the empty circles with 3 and 6 if they do not fully understand the parts and the whole.

ANSWERS Answers for the **Practice** part of the lesson can be found in the *Power Maths* online subscription.

Reflect

WAYS OF WORKING Independent thinking

IN FOCUS Children stretch their thinking on part-whole models in this part of the lesson. In all but one situation the whole will always be the biggest number as it is made up of two parts.

ASSESSMENT CHECKPOINT Have a show of hands to see who thinks the **Reflect** statement is true and who thinks it is false. Can children explain their reasons? Most children are likely to say that it is true unless they are thinking of the number 1; this is the only case where the whole is not bigger than the parts.

ANSWERS Answers for the **Reflect** part of the lesson can be found in the *Power Maths* online subscription.

After the lesson ⏸

- Can children draw an empty part-whole model and explain how they work?
- Were children able to find all the possible part-whole models for a specific number? Did they include examples with zero?

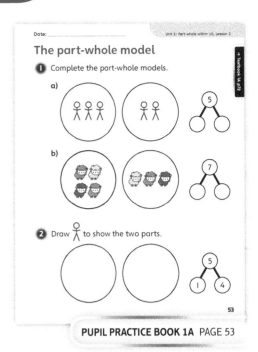

PUPIL PRACTICE BOOK 1A PAGE 53

Date: _____

Unit 2: Part-whole within 10, Lesson 2

The part-whole model

① Complete the part-whole models.

a)

b)

② Draw 👤 to show the two parts.

53

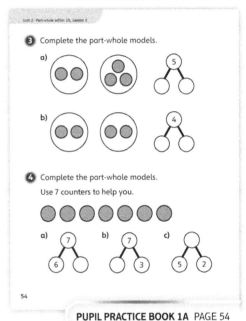

PUPIL PRACTICE BOOK 1A PAGE 54

Unit 2: Part-whole within 10, Lesson 2

③ Complete the part-whole models.

a)

b)

④ Complete the part-whole models.
Use 7 counters to help you.

a) b) c)

54

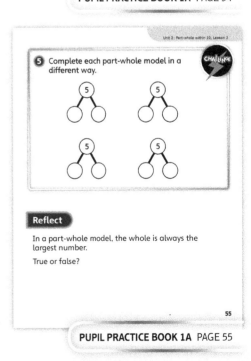

PUPIL PRACTICE BOOK 1A PAGE 55

Unit 2: Part-whole within 10, Lesson 2

⑤ Complete each part-whole model in a different way.

CHALLENGE

Reflect

In a part-whole model, the whole is always the largest number.

True or false?

55

113

Write number sentences

Learning focus

In this lesson, children build on their part-whole knowledge by writing a part-whole as an addition number sentence. For each number in the sentence, they understand what the number represents.

Before you teach

- Consider having some pre-prepared part-whole models for the class for children to write numbers in.
- Are children secure with the idea of parts and the whole?

NATIONAL CURRICULUM LINKS

Number – addition and subtraction

Read, write and interpret mathematical statements involving addition (+), subtraction (–) and equals (=) signs.

Represent and use number bonds and related subtraction facts within 20.

ASSESSING MASTERY

Children can write down an addition number sentence or fact for a given part-whole model and should be able to say what each number represents. They should be able to draw or make a part-whole model from a given number sentence.

COMMON MISCONCEPTIONS

Children may struggle to understand what each number in a number sentence represents. Make sure the number sentence is presented close to the part-whole model. Ask:
- *Can you see what each number represents?*

Children may also put the whole first and do whole + part = part. Ask:
- *What does part add part equal?*
- *Can you use counters to help you see what each number represents?*

STRENGTHENING UNDERSTANDING

Ask children to represent a part-whole model using objects in the classroom. They then write a number sentence under the objects. Give children as much practice as they need to ensure they understand what each part of the number sentence represents. Use the stem sentence from the previous lessons of 'X is a part, Y is a part and Z is the whole' to help them.

GOING DEEPER

Ask children to make their own part-whole models and represent them using real-life objects or cubes. Consider different ways of writing a number sentence, such as whole = part + part. Discuss with children that the parts can be in any order.

KEY LANGUAGE

In lesson: number sentence, plus (+), parts, wholes, part-whole model, equals (=)

Other language to be used by the teacher: represent

STRUCTURES AND REPRESENTATIONS

Part-whole model, ten frame

RESOURCES

Mandatory: counters and cubes

Optional: real-life objects from the classroom such as pencils and pencil pots, string, chalk, hoops

 In the eTextbook of this lesson, you will find interactive links to a selection of teaching tools.

Quick recap 🔍

Provide some counters on a ten frame. Ask children if they can work out how many counters are on the ten frame without counting them. Show different representations of the same number on the ten frame.

Discover

Write number sentences

WAYS OF WORKING Pair work

ASK

- Question ① a): *Can you draw a picture to represent the pencils?*
- Question ① a): *Can you draw a part-whole model to represent the pencils?*
- Question ① a): *Is there more than one way of drawing the part-whole model?*
- Question ① b): *Can you see where the numbers 2, 3 and 5 might be in the diagram and what they represent?*

IN FOCUS Question ① b) introduces children to the idea of linking parts and wholes to a number sentence. Children will also link the words 'equal to' and 'plus' to the correct signs.

PRACTICAL TIPS Use pencils and pencil pots (cups) to recreate the **Discover** scenario at the front of the classroom. For extra practice, ask children to use their own pencils and pots on their tables to recreate the problem.

ANSWERS

Question ① a):

Question ① b): 2 + 3 = 5
              2 is a part.
              3 is a part.
              5 is the whole.

Discover

Tia   Anya

① a) Draw a part-whole model to show the pencils.

   b) Tia writes this number sentence.

       2 + 3 = 5

   What does each number show?

> + means plus
> = means is equal to

76

PUPIL TEXTBOOK 1A PAGE 76

Share

WAYS OF WORKING Whole class teacher led

ASK

- Question ① a): *Does it matter which circles the 2 and the 3 go in?*
- Question ① a): *Does it matter which circle the 5 goes in?*
- Question ① a): *What does each number in the part-whole model represent?*
- Question ① b): *What does each number in the number sentence represent?*

IN FOCUS Children need to see the journey from the concrete through to the pictorial and onto the abstract calculation. Questions ① a) and ① b) link this learning together, helping children embed what the + and = signs mean. If children know what the numbers in each part-whole model represent for the groups, then they should be able to link it to what each number in the number sentences represents. They should use the language 'plus' and 'is equal to'.

Share

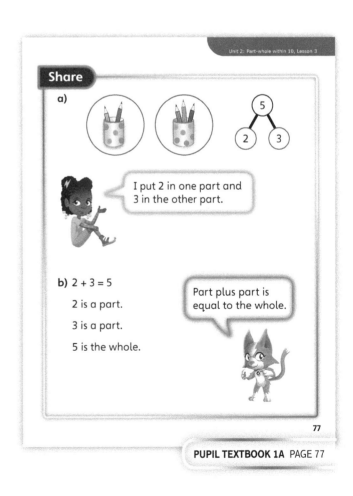

a)

I put 2 in one part and 3 in the other part.

b) 2 + 3 = 5

   2 is a part.

   3 is a part.

   5 is the whole.

Part plus part is equal to the whole.

77

PUPIL TEXTBOOK 1A PAGE 77

Think together

Think together

WAYS OF WORKING Whole class teacher led (I do, We do, You do)

ASK

- Question **1**: *What can you see? Can you see the two parts? What do you think the whole looks like? What are the missing numbers?*
- Question **2**: *Which circle represents the whole? Which circles represent the parts? How do you know that there is a mistake? What should the correct answer be?*

IN FOCUS Question **1** focuses on making sure children know what each number in a number sentence represents. Provide some additional examples if needed. Question **2** asks children to look at an example that is incorrect. They should look at each part and check the part-whole to find which part is incorrect. Ask children to draw their own part-whole models for a number sentence you give them. This will strengthen their understanding.

STRENGTHEN Encourage children to use practical equipment throughout. They should use a variety of resources – some that match the pictures, such as pencils, and others that represent the pictures, such as cubes or counters. You may want to give children some more examples similar to question **1**.

DEEPEN Encourage children to find multiple answers to the questions. Question **3** asks them to break up 7 into parts in different ways. How can the children record all their answers? You may want them to do the same for different numbers.

ASSESSMENT CHECKPOINT By the end of this section, ensure that children can confidently and accurately write a number sentence for a given part-whole model.

ANSWERS

Question **1** a):

Question **1** b): $6 + 3 = 9$

Question **2**: $6 + 2 = 8$.
8 should be the whole (the top number) and 6 and 2 the parts. Hence $8 = 6 + 2$.

Question **3** a): Possible groups are: 1 and 6, 2 and 5, 3 and 4, in any order.

Question **3** b): A completed part-whole model that represents the child's groups from question **3** a).

Question **3** c): A number sentence that represents the child's groups from question **3** a), such as $7 = 1 + 6$ or $7 = 6 + 1$.

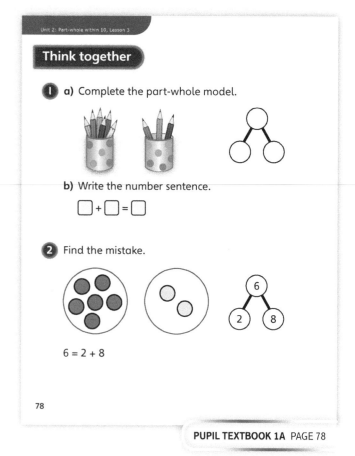

1 a) Complete the part-whole model.

b) Write the number sentence.

$\square + \square = \square$

2 Find the mistake.

$6 = 2 + 8$

78

PUPIL TEXTBOOK 1A PAGE 78

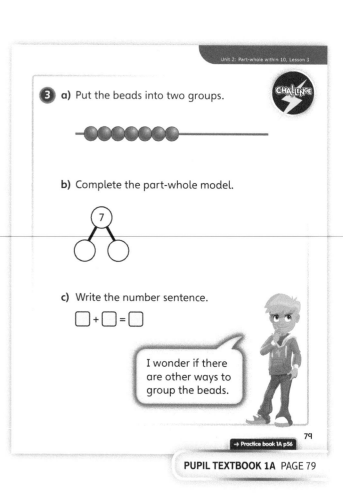

3 a) Put the beads into two groups.

CHALLENGE

b) Complete the part-whole model.

c) Write the number sentence.

$\square + \square = \square$

I wonder if there are other ways to group the beads.

79

→ Practice book 1A p56

PUPIL TEXTBOOK 1A PAGE 79

Practice

WAYS OF WORKING Independent thinking

IN FOCUS The questions are presented in a variety of ways to check whether children are confident with the concepts of parts and wholes and linking these with abstract number sentences. Question ❶ provides pictorial support in completing part-whole models and questions ❷ to ❺ link part-whole models to abstract number sentences. Children need to be able to use the plus (+) and equal to (=) signs and understand that number sentences can be written in different orders.

STRENGTHEN Children can use concrete materials for each question to support their understanding. All of the questions could also be set up in the classroom using large circles of string to represent the part-whole model, or outside in the playground using chalk circles or hoops. Ask children to draw their own part-whole models for a given number sentence.

DEEPEN For question ❺, encourage children to think about how they record their results when a question has multiple answers. Can they see a pattern between the number they start with and the number of ways it can be partitioned?

THINK DIFFERENTLY Question ❹ asks children to partition 5 in different ways and to show this in part-whole models and number sentences.

ASSESSMENT CHECKPOINT The first few questions will help you understand whether children have grasped the purpose of a number sentence. Can they write a number sentence for a given part-whole model and know what each number represents? Question ❹ checks that children are able to find different ways of partitioning 5. Check how they have completed this. Have they filled in the part-whole models randomly or been systematic in their approach? Have they understood that 5 + 0 and 0 + 5 are also correct solutions?

ANSWERS Answers for the **Practice** part of the lesson can be found in the *Power Maths* online subscription.

Reflect

WAYS OF WORKING Pair work

IN FOCUS Children write down their own number sentence. Encourage them to draw a part-whole model to represent this number sentence and explain to their partner what each number represents. Ask their partner to check if they agree with them.

ASSESSMENT CHECKPOINT Do children know what each of the numbers in the number sentence represents? Do they know what the numbers in the part-whole model represent?

ANSWERS Answers for the **Reflect** part of the lesson can be found in the *Power Maths* online subscription.

After the lesson ⏸

- Are children confident using the + and = signs?
- Do they understand which part of the number sentence represents the whole and which parts represent the parts?

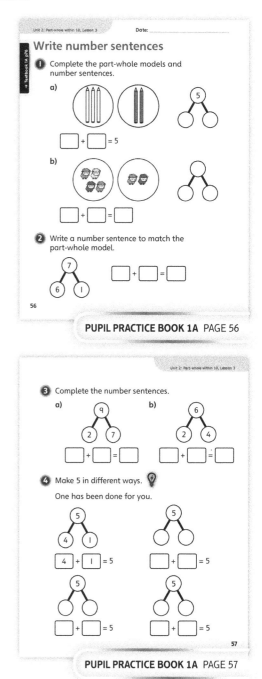

PUPIL PRACTICE BOOK 1A PAGE 56

PUPIL PRACTICE BOOK 1A PAGE 57

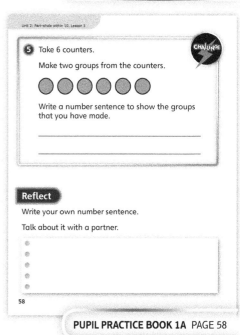

PUPIL PRACTICE BOOK 1A PAGE 58

Fact families – addition facts

Learning focus

In this lesson, children consolidate their learning on number sentences. They explore related facts, understand the term 'fact family' and are able to write one down for a part-whole model.

Before you teach

- Do children understand what the + and = signs mean?
- Do children understand that numbers can be partitioned in different ways?

NATIONAL CURRICULUM LINKS

Number – addition and subtraction

Read, write and interpret mathematical statements involving addition (+), subtraction (–) and equals (=) signs.

Represent and use number bonds and related subtraction facts within 20.

ASSESSING MASTERY

Children can write all the possible number sentences for any part-whole model and can make their own number sentences. Children can recognise that there are four different ways of writing an addition sentence.

COMMON MISCONCEPTIONS

Children may get their numbers mixed up and write more number sentences than there are. For example, for 4 + 2 = 6 they may be tempted to write 4 + 6 = 2 or 2 + 6 = 4. Ask:

- *Can you prove that 4 + 6 = 2 is correct using cubes?*
- *What have you done wrong?*
- *What is 4 + 6 actually equal to? How can you tell?*

It is too soon to link this to subtraction sentences, so ensure that you are only discussing what the whole is with the children to prove why their sentence is incorrect.

STRENGTHENING UNDERSTANDING

Give children access to blocks, cubes or counters and ask them to use these resources to represent simple number sentences. Ask them to move the resources around, as illustrated below, to prove that the = sign can go in different places and the number sentence remains correct. For example:

 + =   = +

3 + 1 = 4   4 = 1 + 3

GOING DEEPER

To deepen learning, children could look at partitioning numbers into three parts instead of two. For example, they could write 7 = 1 + 2 + 4 or 2 + 2 + 3 = 7. Ask children to partition different numbers to 10 into three parts.

KEY LANGUAGE

In lesson: groups, number sentence, same, different, fact family

Other language to be used by the teacher: part-whole model, partition, part, whole, equal to, plus

STRUCTURES AND REPRESENTATIONS

Part-whole model

RESOURCES

Mandatory: cubes, blocks, counters

Optional: cups or glasses of juice or water, plastic food and flowers, hoops, teddy bears, toys, etc.

 In the eTextbook of this lesson, you will find interactive links to a selection of teaching tools.

Quick recap

Ask children to represent a simple number sentence that they might have come across in the previous lesson by drawing or using concrete objects. For example, 3 + 1 = 4. Ask them to point out what each number represents.

Discover

Unit 2: Part-whole within 10, Lesson 4

Fact families – addition facts

WAYS OF WORKING Pair work

ASK

- Question ① a): *What do you have on the table?*
- Question ① a): *Can you see the two groups?*
- Question ① a): *What do you think the numbers in the part-whole model should be?*
- Question ① b): *Is there just one way of writing down your answer?*
- Question ① b): *Is there any way you can check your answer?*

IN FOCUS Questions ① a) and ① b) reinforce children's understanding of forming part-whole models and then forming number sentences. They begin to see that a number sentence can be written in different ways.

PRACTICAL TIPS Use cups or glasses of juice and water to recreate the scenario in the classroom.

ANSWERS

Question ① a):

Question ① b): 2 + 4 = 6
4 + 2 = 6

Discover

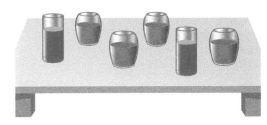

① a) Complete the part-whole model to show the number of glasses.

b) Write the number sentence.

Did you write the same as your partner?

80

Share

WAYS OF WORKING Whole class teacher led

ASK

- Question ① a): *How many different ways has your group written it?*
- Question ① a): *Do you think there are any more ways?*
- Question ① b): *Compare your number sentence with the rest of the group. Have you all written it the same way?*
- Question ① b): *Can you see why it is called a fact family?*

IN FOCUS Children share their ideas and come to the conclusion that there are two (or some may see four) different addition sentences that can be written about one part-whole diagram. Children should have given one of the two possible part-whole models here. Some children may have put the 4 first. Discuss with children that the part-wholes represent the same thing. For each part-whole we can write down a number sentence. Explain that both these sentences are correct.

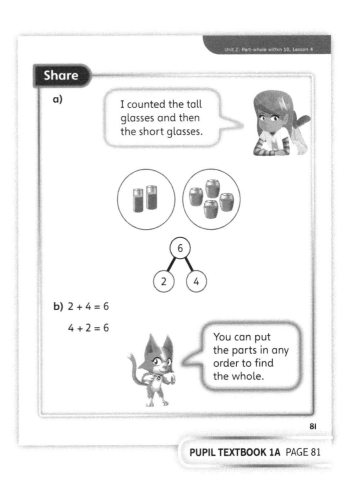

119

Think together

WAYS OF WORKING Whole class teacher led (I do, We do, You do)

ASK

- Question ❶: *What parts can you see? Can you put the numbers in a part-whole in a different way? How many number sentences can you write? Do you need to draw a different part-whole to write both number sentences?*
- Question ❷: *Can you see the two groups of flowers? Can you write both number sentences for the flowers?*

IN FOCUS In question ❶, children put numbers in the part-whole and write a number sentence. Discuss the two different ways they may have done this and the two number sentences that they may have got. Because the parts are not labelled, some children should come up with each of the sentences. Question ❷ then builds on this by asking them to try and write two sentences without necessarily drawing the part-whole. By the end of question ❷ children should be able to write down two sentences confidently and correctly.

STRENGTHEN Throughout, children can use concrete materials to represent the objects. For example, different coloured counters to represent the two groups of flowers. Work closely with children to explain that the parts can be written in either order.

DEEPEN In question ❸, children are given the opportunity to write two number sentences for the part-whole. They see that the whole can be written at the start as well as at the end. You could further challenge children to draw a picture of some flowers that could be partitioned in three ways.

ASSESSMENT CHECKPOINT By the end of this section, children should be able to write confidently and accurately at least two addition sentences for a part-whole model.

ANSWERS

Question ❶: The parts could be 1 and 4 or 2 and 3, in any order.
The number sentence should reflect the part-whole model.
$1 + 4 = 5$
$4 + 1 = 5$
$2 + 3 = 5$
$3 + 2 = 5$

Question ❷: $5 + 2 = 7$
$2 + 5 = 7$

Question ❸: $1 + 5 = 6$
$5 + 1 = 6$
$6 = 1 + 5$
$6 = 5 + 1$

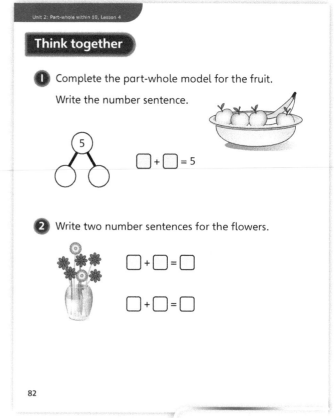

PUPIL TEXTBOOK 1A PAGE 82

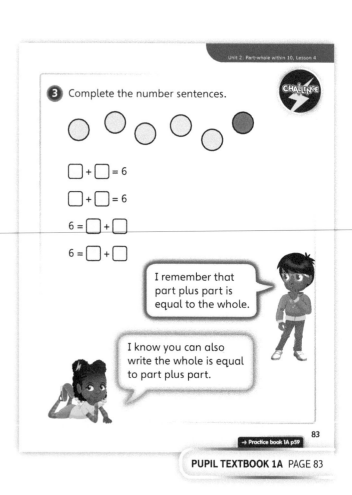

PUPIL TEXTBOOK 1A PAGE 83

Practice

WAYS OF WORKING Independent thinking

IN FOCUS In questions ❶ and ❷, children are asked to write number sentences from pictorial representations. The scaffolding is reduced in questions ❸ and ❺, where children are asked to write number sentences from part-whole models. They consolidate the idea that for every addition fact they know, they can write down three more.

STRENGTHEN Give children a number sentence and ask them to draw a picture of what this sentence could represent. Make sure that they are given number sentences written in a variety of ways.

DEEPEN To deepen understanding, children could make their own questions for their partner to answer. They could make questions that start with a picture, such as three cats and six dogs, and ask their partner to complete a part-whole model and number sentences to match the picture. To further extend learning, ask children to write some questions that start with a part-whole model, and their partner could then draw a picture and write the matching number sentences.

THINK DIFFERENTLY Question ❹ asks children to complete a part-whole model and write the corresponding number sentences from a pictorial representation. Neither the parts or whole are shown. Children should realise that the number of bananas is one part, the number of apples is another part, and the total number of pieces of fruit is the whole.

ASSESSMENT CHECKPOINT

Question ❸ reinforces that the addition can be reversed and still give the same answer and that the total can come first, with the addition after the equals sign. Question ❺ requires children to write the addition and equals signs for the first time. Check that they recognise that the different-shaped missing box means they should write a sign there.

ANSWERS Answers for the **Practice** part of the lesson can be found in the *Power Maths* online subscription.

Reflect

WAYS OF WORKING Independent thinking

IN FOCUS This activity ensures that children have made the link between a part-whole model and the four addition sentences that go with it.

ASSESSMENT CHECKPOINT Check that children have drawn a part-whole model that makes sense. Have they put the numbers in the correct place? Check that they have written all four sentences correctly.

ANSWERS Answers for the **Reflect** part of the lesson can be found in the *Power Maths* online subscription.

After the lesson ❚❚

- Are children confident writing four addition number sentences for one part-whole model?
- Are children confident at various starting points? For example, can they start with the number sentence and work backwards to draw a part-whole model? Can they start with a part-whole model, then draw a diagram to represent the model (for example, they might draw 4 cherries in a bowl and 2 cherries in another bowl, equalling 6 cherries in a third bowl). Can they then write four number sentences?
- How can you build this into future learning?

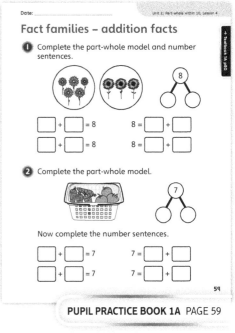

PUPIL PRACTICE BOOK 1A PAGE 59

PUPIL PRACTICE BOOK 1A PAGE 60

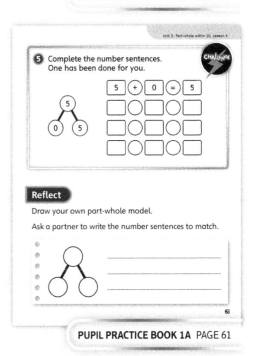

PUPIL PRACTICE BOOK 1A PAGE 61

121

Number bonds

Learning focus

In this lesson, children start to look at number bonds to 10. This lesson is not about the systematic approach to finding number bonds, but more about understanding the term 'bond' and being able to write down number bonds.

Before you teach

- Check children know how to draw a part-whole model for a given number sentence.
- Check children know what each number in a number sentence represents.

NATIONAL CURRICULUM LINKS

Number – addition and subtraction

Represent and use number bonds and related subtraction facts within 20.

ASSESSING MASTERY

Children can write down a number bond to numbers within 10, supported by an image or representation, and should start to recognise some number bonds to 10 without having to count on their fingers. This work on number bonds should be reinforced through **Power Ups** or other activities across the year.

COMMON MISCONCEPTIONS

Children may need to count on their fingers to work out certain bonds. Therefore, they may make the mistake of starting to count from the first number rather than starting with the number after. Ask:
- *What is your starting number?*
- *What number comes next?*

STRENGTHENING UNDERSTANDING

Use concrete objects to make number bonds. You may want to use counters on a ten frame, a bead string or rekenrek to help them understand. Encourage children to subitise small numbers rather than counting.

GOING DEEPER

For children already confident with number bonds to 5 or 10, ask them to find all the bonds to a certain number. Look for the approach they take and whether or not it is systematic.

KEY LANGUAGE

In lesson: number bond

Other language to be used by the teacher: represent, objects

RESOURCES

Mandatory: counters or cubes

Optional: bead strings, rekenreks, ten frames

 In the eTextbook of this lesson, you will find interactive links to a selection of teaching tools.

Quick recap

Ask children to put six doubled-sided counters in a cup. Tap the counters out. Play a game where a child with four red counters wins. Keep playing the game and changing the number. (If you don't have access to double-sided counters, you could stick a spot on one side of each counter.)

Discover

WAYS OF WORKING Pair work

ASK

- Question ❶ a): *Can you make the tower of cubes that Jack has made?*
- Question ❶ a): *How many cubes did you use?*
- Question ❶ a): *What has Jack done with this tower? Can you do that?*
- Question ❶ b): *Can you see Jack has made two parts? How many are in each part?*
- Question ❶ b): *Could you break Jack's tower in different ways?*
- Question ❶ b): *Can you write a number sentence to show a bond to 10?*

IN FOCUS This practical activity that children will replicate gets them to see how numbers can be broken into parts and two numbers can come together to make a whole.

PRACTICAL TIPS Children use five cubes to recreate the **Discover** scenario.

ANSWERS

Question ❶ a): Jack broke his tower into two parts.

Question ❶ b): $5 = 2 + 3$
$2 + 3 = 5$

Number bonds

Discover

Before · After

❶ a) What did Jack do to his tower of cubes?

b) Write a number sentence to show this.

84

PUPIL TEXTBOOK 1A PAGE 84

Share

WAYS OF WORKING Whole class teacher led

ASK

- Question ❶ a): *Can you make the tower? Can you do what Jack has done?*
- Question ❶ a): *What number does the tower represent?*
- Question ❶ b): *Can you write down a number sentence for this?*
- Question ❶ b): *What do we call this number sentence?*
- Question ❶ b): *Do you know any other number bonds?*

IN FOCUS In question ❶ a), it is important to go through what Jack has done with the children as a whole class. Ask them to replicate what Jack has done using cubes. You should gesture the breaking apart and coming together.

Discuss what Sparks says, as it is the key learning of this lesson. Children need to understand that the term 'number bond' shows two numbers that add together to make another number. Use the number sentence 2 + 3 = 5 as an example. Ask children if they know any other number bonds.

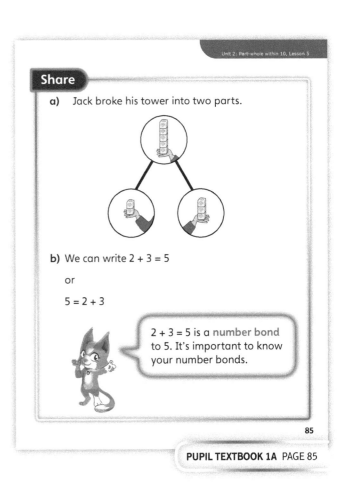

Share

a) Jack broke his tower into two parts.

b) We can write $2 + 3 = 5$

or

$5 = 2 + 3$

$2 + 3 = 5$ is a number bond to 5. It's important to know your number bonds.

85

PUPIL TEXTBOOK 1A PAGE 85

Think together

WAYS OF WORKING Whole class teacher led (I do, We do, You do)

ASK

- Question **1**: *What has Jack done this time? What is the same? What is different?*
- Question **2**: *What number bonds can you see? Where does each number in the bond come from?*
- Question **3**: *How would you approach this question?*

IN FOCUS In question **1**, children go back to Jack's model and create a different number bond. It is essential that children see that different numbers make up the number 5.

Question **2** provides a different way of showing a number bond using different coloured counters. Question **2** allows all children to see how the same number can be made up in different ways. They should be encouraged to subitise.

Throughout this section, you may want to focus on using one representation. For example, for questions **2** and **3**, you may want to just use the towers of cubes if children find it easier.

STRENGTHEN Work closely with children who are struggling. Use cubes and coloured counters to exactly replicate the different questions. Use gesturing to help them understand what to do to break up the towers.

DEEPEN Deepen understanding by asking children to be more systematic with their approach to forming number bonds. Give them a number and ask them if they can find all the number bonds, similar to the **Challenge** example which gives a method to make all the number bonds to 7.

ASSESSMENT CHECKPOINT Children should be able to write down number bonds to a number within 10 using objects and representations to support them. Check that children have been able to write down correct number bonds.

ANSWERS

Question **1**: 4 + 1 = 5 or 1 + 4 = 5

Question **2** a): 3 + 3 = 6

Question **2** b): 4 + 2 = 6 or 2 + 4 = 6

Question **3**: 3 + 4 = 7
4 + 3 = 7
2 + 5 = 7
5 + 2 = 7
1 + 6 = 7
6 + 1 = 7
0 + 7 = 7
7 + 0 = 7

Think together

1 Complete the number bond to 5.

$\square + \square = 5$

2 What number bonds to 6 can you see here?

a)

b)

86

PUPIL TEXTBOOK 1A PAGE 86

CHALLENGE

3 Take seven double-sided counters.

Put them in the circle to make different number bonds to 7.

Write your number bonds.

I wonder how many different number bonds to 7 there are.

87

→ Practice book 1A p62

PUPIL TEXTBOOK 1A PAGE 87

Practice

IN FOCUS Questions ❶ to ❸ ask children to look at number bonds and write them down. Encourage children to say them out loud and point to what each number represents. Encourage children to subitise as opposed to counting. Ideally you want them to see 2 or 3 rather than count each number. Representations have been provided that are common for children.

STRENGTHEN Work closely with children who are struggling. Use cubes and coloured counters to replicate the different questions.

DEEPEN Look at how children approach questions ❹ and ❺. Do they approach the questions randomly or systematically?

ASSESSMENT CHECKPOINT By the end of these questions, children should be more aware of number bonds of particular numbers.

ANSWERS Answers for the **Practice** part of the lesson can be found in the *Power Maths* online subscription.

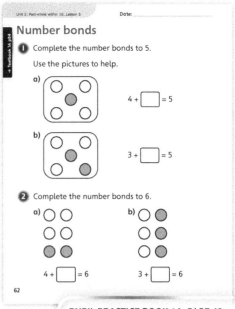

PUPIL PRACTICE BOOK 1A PAGE 62

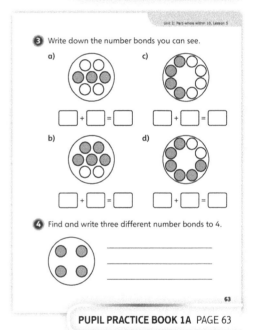

PUPIL PRACTICE BOOK 1A PAGE 63

Reflect

IN FOCUS The purpose of this activity is for children to explore making number bonds with six cubes.

ASSESSMENT CHECKPOINT Check children have understood the word 'bond' and that they can write at least one number bond to 6. Some children may be able to write more than one number bond.

ANSWERS Answers for the **Reflect** part of the lesson can be found in the *Power Maths* online subscription.

After the lesson

- Check if children know some number bonds without having to use images or concrete objects.
- Continue to reinforce number bonds throughout the year.

PUPIL PRACTICE BOOK 1A PAGE 64

Find number bonds

Learning focus

In this lesson children are learning about number bonds within 10. Children learn strategies for organising their thinking and begin to spot patterns.

Before you teach

- Do children understand that 3 + 1 is the same as 1 + 3?
- Do children understand that the equals sign can be put in different places, such as 3 + 1 = 4 and 4 = 3 + 1?

NATIONAL CURRICULUM LINKS

Number – addition and subtraction

Represent and use number bonds and related subtraction facts within 20.

ASSESSING MASTERY

Children can work systematically to find all the number bonds of a number within ten.

COMMON MISCONCEPTIONS

Children may work in a random way and therefore miss or duplicate answers. Children may also forget or disregard number bonds including zero. Ask:

- *Can you think of a different way of working to make sure you do not miss any number bonds?*
- *You have missed one of the number bonds. Can you think what you may have missed?*

STRENGTHENING UNDERSTANDING

Using practical equipment in two different colours, such as counters, cubes or bead strings, will help children see patterns and understand when they have found all of their number bonds. For example, here we can see the number bonds for 4:

●●●● 4 + 0
●●●○ 3 + 1
●●○○ 2 + 2
●○○○ 1 + 3
○○○○ 0 + 4

By changing one block at a time, children can see that they have found all the number bonds.

GOING DEEPER

Children could expand their thinking by working on more complex problems involving number bonds. For example:

If the circle has the same value in each calculation, what must the value of the square be?

○ + 5 = 7
□ + ○ = 5

KEY LANGUAGE

In lesson: how many, same, different, number sentence

Other language to be used by the teacher: number bond, part-whole model, partition, part, whole, equal to, sequence

STRUCTURES AND REPRESENTATIONS

Ten frame, five frame, bead string

RESOURCES

Mandatory: ten frame, five frame, bead string

Optional: cubes, counters, rekenrek

 In the eTextbook of this lesson, you will find interactive links to a selection of teaching tools.

Quick recap

Ask children if they know any number bonds. Put some number bonds to 5 on the board and ask children to give the answer without working them out.

Discover

Find number bonds

WAYS OF WORKING Pair work

ASK

- Question **1** a): *How many different colours are used?*
- Question **1** a): *How many red counters are there and how many yellow?*
- Question **1** a): *How many are there in total?*
- Question **1** b): *How many different ways can you find to fill the five frame?*

IN FOCUS Children experiment with different ways to make five.

PRACTICAL TIPS For question **1** a), ask children to work in small groups to recreate the scenario with counters and five frames.

ANSWERS

Question **1** a): $2 + 3 = 5$

Question **1** b): $3 + 2 = 5$
$1 + 4 = 5$
$4 + 1 = 5$
$0 + 5 = 5$
$5 + 0 = 5$

Discover

I have used two colours to fill my five frame.

1 **a)** Write the number sentence for the counters.

b) Use double-sided counters.

Find two more ways to fill a five frame.

88

PUPIL TEXTBOOK 1A PAGE 88

Share

WAYS OF WORKING Whole class teacher led (I do, We do, You do)

ASK

- Question **1** b): *Did everyone complete the activity in the same way?*
- Question **1** b): *How do we know we have found all the ways?*
- Question **1** b): *Should we start with all the same-coloured counters, then swap a counter one by one?*

IN FOCUS Children share ideas and begin to think of strategies to ensure they have found all the different ways of filling in the five frame.

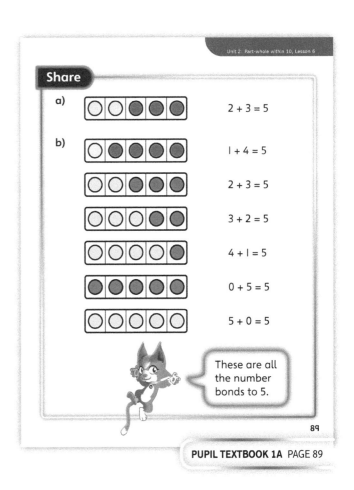

Share

a) $2 + 3 = 5$

b) $1 + 4 = 5$

$2 + 3 = 5$

$3 + 2 = 5$

$4 + 1 = 5$

$0 + 5 = 5$

$5 + 0 = 5$

These are all the number bonds to 5.

89

PUPIL TEXTBOOK 1A PAGE 89

Think together

Whole class teacher led (I do, We do, You do)

ASK

- Question **1**: *What can you see? What numbers do you think make up 4?*
- Question **2**: *How can we find all the number bonds to 4? How can you be sure you do not miss any out? Explain your method to a partner.*
- Question **3**: *Do you think you have found all the ways of making 6? How can you check?*

IN FOCUS Children focus on being systematic with their method for finding number bonds using a variety of representations. Questions **1** and **2** ask children to find all the number bonds to 4.

STRENGTHEN These questions could be replicated practically using children with and without objects in the classroom. The children could use counters, cubes or a bead string to illustrate what is happening. Equally you may want to use a tower of cubes similar to the approach children took in the previous lesson.

DEEPEN Start to look at number bonds to 8 or 9 using a ten frame. Can they find all the number bonds?

Extend learning by encouraging children to experiment with partitioning numbers in three ways. Encourage them to think of ways of making sure they have found all possible number bond combinations.

ASSESSMENT CHECKPOINT Questions **1** and **2** should highlight whether children can systematically find the number bonds to a particular number.

ANSWERS

Question **1** a): $4 + 0 = 4$

Question **1** b): $3 + 1 = 4$

Question **2** a): $2 + 2 = 4$

Question **2** b) and c):

      Additions should reflect their drawings:

      $1 + 3 = 4$

      $3 + 1 = 4$

      $4 + 0 = 4$

      $0 + 4 = 4$

Question **3**: $6 = 3 + \mathbf{3}$

      $6 = \mathbf{4} + 2$

      $6 = \mathbf{5} + 1$

PUPIL TEXTBOOK 1A PAGE 90

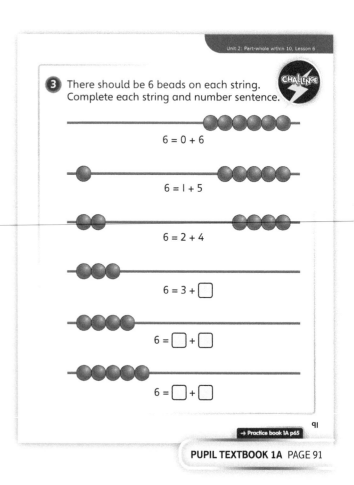

PUPIL TEXTBOOK 1A PAGE 91

Practice

WAYS OF WORKING Independent thinking

IN FOCUS Children practise finding number bonds independently. Question ① asks children to practise number bonds to 4, question ② asks children to practise number bonds to 7, question ③ asks children to practise numbers bonds to 6, question ④ asks children to practise number bonds to 9, and question ⑤ asks children to practise number bonds to 8. Children should look for patterns to help them complete the additions.

STRENGTHEN Encourage children to represent the pictures with the actual equipment throughout. For question ④, allow children to choose which equipment they would prefer to use to represent the number sentences.

DEEPEN Extend learning by giving children some open-ended problems to solve such as:

$$\Box + \Box + \Box = 9$$

Ask: *Which numbers could you write in these boxes to make the answer 9? Could the numbers all be the same? What if the numbers all have to be different? How many different answers can you find?*

Ask children to create some of their own problems similar to those in question ④.

ASSESSMENT CHECKPOINT Question ② highlights whether children can see the patterns. Question ⑤ checks whether children can find solutions systematically and record additions correctly.

ANSWERS Answers for the **Practice** part of the lesson can be found in the *Power Maths* online subscription.

Reflect

WAYS OF WORKING Independent thinking

IN FOCUS Children prove that they have understood the lesson by drawing beads to show number bonds. You should expect to see children being organised in their approach.

ASSESSMENT CHECKPOINT Check that children have drawn their beads systematically, not randomly.

ANSWERS Answers for the **Reflect** part of the lesson can be found in the *Power Maths* online subscription.

After the lesson

- Have children been systematic with their approach to finding number bonds?
- Are they able to show number bonds using pictures and a number sentence?
- Have children made use of zero throughout the lesson?

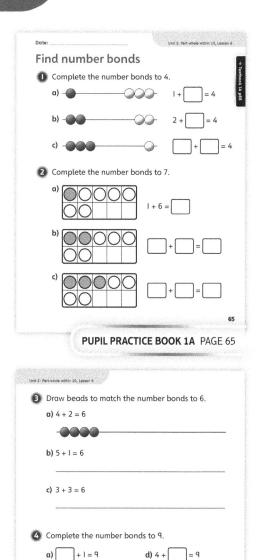

PUPIL PRACTICE BOOK 1A PAGE 65

PUPIL PRACTICE BOOK 1A PAGE 66

PUPIL PRACTICE BOOK 1A PAGE 67

Number bonds to 10

Learning focus

In this lesson, children will find and represent number bonds to 10. The lesson builds on the previous one, on finding a missing part.

Before you teach

- Based on teaching of part-whole models earlier in the unit are there any difficulties or misconceptions that need to be tackled before starting this lesson?
- Using your knowledge of how children have worked with number bonds to 10 in previous lessons, are there any adaptations or links you can make to this lesson?
- How could you prompt children to use their fingers or cubes to help them during this lesson?

NATIONAL CURRICULUM LINKS

Number – addition and subtraction

Represent and use number bonds and related subtraction facts within 20.

ASSESSING MASTERY

Children can use instant recall of number bonds to 10, and represent them in a ten frame and a part-whole model. Children can use this knowledge to answer missing-number problems without having to count on.

COMMON MISCONCEPTIONS

Children may not make the link between instant recall of number bonds to 10 and solving missing-number problems. They may revert to counting on, even if they know the answer. If children revert back to working out the number bond, use different mathematical structures to represent calculations. Ask:
- *Can you highlight the numbers that are the same?*

STRENGTHENING UNDERSTANDING

You can reinforce learning by encouraging children to work systematically to find all of the number bonds to 10. Suggest that they show this working on a resource such as a bead string.

GOING DEEPER

You can deepen learning by encouraging children to generate their own calculations based on their knowledge of number bonds to 10. Scaffolds, such as $\boxed{} + \boxed{} = 10$, could be given.

When confident, children can create their own missing-number calculations in which the whole is always 10. For example, $8 + \boxed{} = 10$. Remember to explore $0 + 10$, as this is a number bond children should know.

KEY LANGUAGE

In lesson: altogether, add / added / adding / addition, plus, '+', in total, sum, number bonds

Other language to be used by the teacher: whole, part

STRUCTURES AND REPRESENTATIONS

Part-whole model, ten frame

RESOURCES

Mandatory: blank part-whole model, blank ten frame, counters or cubes
Optional: bead string

 In the eTextbook of this lesson, you will find interactive links to the following teaching tools:

Quick recap

Give children a tower of seven cubes. Ask them to use the cubes to find all the number bonds to 7. How many can they find in 3 minutes?

Discover

WAYS OF WORKING Pair work

ASK

- Question ① a): *How many cans are upright?*
- Question ① a): *How many cans are on their side?*
- Question ① a): *How many cans are there in total?*

IN FOCUS In this part of the lesson, children are exposed to a real-life practical context surrounding number bonds to 10. Question ① introduces a concrete story as an illustration of the calculation. Focus on the picture. Ask: *What do you see? What do you think the problem could be?*

PRACTICAL TIPS Recreate the **Discover** scenario using 3D shapes or other real-life objects.

STRENGTHEN Strengthen children's understanding by acting out the story in the classroom, stacking 10 objects on a classroom table. Encourage children to play with the objects and record the number bonds they find.

ANSWERS

Question ① a): There are 7 upright cans.
There are 3 cans on their side.
There are 10 cans altogether

Question ① b): 5 + 5 = 10

Number bonds to 10

Discover

Poppy   Kendi

① **a)** Look at the number sentence for Kendi.

$7 + 3 = 10$

What does each number show?

b) Write a number sentence for Poppy.

92

PUPIL TEXTBOOK 1A PAGE 92

Share

WAYS OF WORKING Whole class teacher led

ASK

- Question ①: *What role does 10 have in these calculations?*
- Question ① a): *What can we use to find bonds to 10?*
- Question ① a): *Why is a ten frame useful?*
- Question ① b): *On a ten frame, what numbers can you see easily?*

IN FOCUS This part of the lesson encourages children to share different strategies for finding number bonds to 10, based on structures and representations used in previous lessons.

STRENGTHEN If children struggle to recall different representations, strengthen their understanding by talking together about the different ways you can find the answer to 3 + 7. Ask: *What is the same about these methods, and what is different? Does the whole remain the same? Where are the parts 3 and 7? Does it matter which way around they go?*

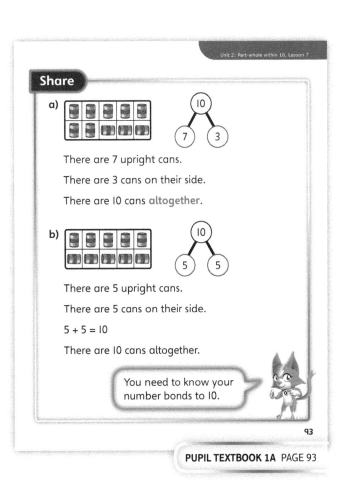

Share

a)

There are 7 upright cans.

There are 3 cans on their side.

There are 10 cans altogether.

b)

There are 5 upright cans.

There are 5 cans on their side.

$5 + 5 = 10$

There are 10 cans altogether.

You need to know your number bonds to 10.

93

PUPIL TEXTBOOK 1A PAGE 93

Think together

WAYS OF WORKING Whole class teacher led (I do, We do, You do)

ASK

- Question ❶: *Can you point to how many cans are on their sides? What number is shown in the part-whole model – the number of cans that are upright or the number of cans that are on their sides?*
- Question ❷: *What two parts of the part-whole model do you need to fill in?*

IN FOCUS In question ❶, 6 cans are shown upright, and 4 are shown on their sides. This illustration reinforces the number bond 6 + 4 = 10. Children can work out the answer to the calculation by using their knowledge of number bonds or by referring to the artwork.

Draw attention to what Astrid points out after question ❸, about using hands as a way to remember 5 + 5 = 10. Alternatively, children could use cubes or counters to show this.

In question ❸, there is systematic progression in finding the number bonds to 10. This can be reinforced by working concretely with 10 objects and moving one at a time.

STRENGTHEN Strengthen children's understanding of how to tackle question ❷ by modelling looking at the picture to find which cans are lying down.

DEEPEN In question ❸, six number bonds to 10 are found. Deepen children's understanding by explaining how they can use these bonds to work out other bonds to 10. For example, if they know that 9 + 1 = 10, they also know that 1 + 9 = 10.

ASSESSMENT CHECKPOINT In question ❸, there are blank ten frames to fill in. Up to this point, children have been presented with filled-in ten frames. Assess how children draw on separate parts for themselves. Will they use different-coloured pens, different-sized dots or another method? Point out what Dexter says about working in order and ask children to explain what this means in relation to what they have just done.

ANSWERS

Question ❶: 6 + 4 = 10

Question ❷: 8 + 2 = 10

Question ❸: 8 + **2** = 10   7 + **3** = 10
          6 + **4** = 10   5 + **5** = 10

Think together

❶ Complete the part-whole model and the number sentence.

6 + ☐ = 10

❷ Complete the part-whole model and the number sentence.

☐ + ☐ = 10

94

PUPIL TEXTBOOK 1A PAGE 94

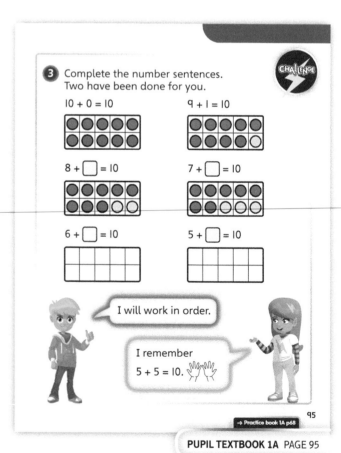

❸ Complete the number sentences. Two have been done for you.

10 + 0 = 10   9 + 1 = 10

8 + ☐ = 10   7 + ☐ = 10

6 + ☐ = 10   5 + ☐ = 10

I will work in order.

I remember 5 + 5 = 10.

95

→ Practice book 1A p68

PUPIL TEXTBOOK 1A PAGE 95

Practice

WAYS OF WORKING Independent thinking

IN FOCUS Questions **1** and **2** provide children with practice of their number bonds to 10. Some children may need to draw the objects to complete 10, others may not. Ask children what the numbers in the number sentences represent.

STRENGTHEN In question **1**, encourage children to draw the objects to make 10. Strengthen children's understanding by inviting them to show the different number bonds using different physical resources, for example, with a bead string.

DEEPEN Deepen understanding of question **4** by challenging children to write number sentences for each pair of numbers.

THINK DIFFERENTLY Question **3** involves adding 0. Reinforce understanding that adding 0 means adding nothing and the original number has not changed.

ASSESSMENT CHECKPOINT Children should be developing their confidence with number bonds to 10 as the lesson progresses. Look out for any children who count the number of circles needed to fill the ten frame, or who rely on their fingers to check or count, rather than trusting their knowledge of number bonds.

ANSWERS Answers for the **Practice** part of the lesson can be found in the *Power Maths* online subscription.

Reflect

WAYS OF WORKING Whole class

IN FOCUS The **Reflect** question reinforces different strategies children can use to remember number bonds to 10. Ask children to remember what other structures they have used to show these bonds.

ASSESSMENT CHECKPOINT Assess to what extent children are using instant recall. Do they have to use their fingers or cubes to help them work out the bonds?

ANSWERS Answers for the **Reflect** part of the lesson can be found in the *Power Maths* online subscription.

After the lesson ⏸

- Are children secure enough in number bonds to 10 to move on to finding addition facts in the next lesson, or is further reinforcement required?
- Did children recognise when their calculation strategies were inefficient, and were they able to modify their strategies based on your intervention?

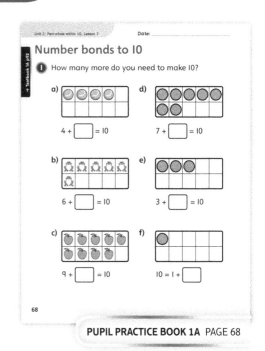

PUPIL PRACTICE BOOK 1A PAGE 68

PUPIL PRACTICE BOOK 1A PAGE 69

PUPIL PRACTICE BOOK 1A PAGE 70

End of unit check

Don't forget the unit assessment grid in your *Power Maths* online subscription.

WAYS OF WORKING Group work adult led

IN FOCUS

- Questions **4** and **5** include a number bond where one of the numbers is 0.
- Questions **2** and **4** require a secure understanding of the = sign.

Think!

WAYS OF WORKING Pair work or small groups

IN FOCUS This activity gives children more than one thing to think about: they need to think about the parts and the whole whilst also considering which numbers will fit where. Check how children approach the problem. How confident are they, faced with an unfamiliar problem without an immediate answer? Do they use a systematic approach or try numbers at random? Challenge children who tackle this problem well by asking them to find more solutions.

Draw children's attention to the key vocabulary at the bottom of the **My journal** page.

Encourage children to think through or discuss possible numbers for all three part-whole models before writing their answer in **My journal**.

ANSWERS AND COMMENTARY Children who have mastered the concepts of this unit will recognise the part-whole model and be able to explain what each number represents. They can use the part-whole model to write the four addition facts that the model represents and they are beginning to learn the number bonds to ten.

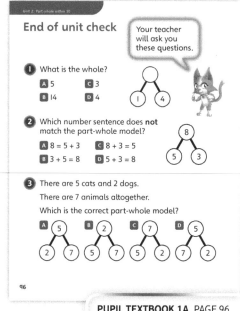

PUPIL TEXTBOOK 1A PAGE 96

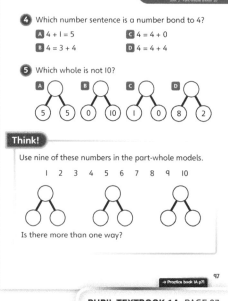

PUPIL TEXTBOOK 1A PAGE 97

Q	A	WRONG ANSWERS AND MISCONCEPTIONS	STRENGTHENING UNDERSTANDING
1	A	C suggests that children think they need to subtract; they may not yet have a secure understanding of how the part-whole model works, whereas B suggests that children think they just need to choose the largest number.	Allow children to use cubes to check and prove their answers as they work.
2	C	A suggests that children think the = sign should always come at the end of the number sentence.	
3	C	Any wrong answer indicates that children do not have a secure understanding of the total or whole.	
4	C	Any wrong answer indicates that children do not recognise $4 = 4 + 0$ as a correct number fact because it includes zero.	
5	C	B suggests the children do not recognise that $10 = 0 + 10$ is a correct number fact. A or D suggest work on addition facts is needed.	

My journal

WAYS OF WORKING Independent thinking

ANSWERS AND COMMENTARY

There are 10 unique solutions:
$10 = 9 + 1, 8 = 6 + 2, 7 = 4 + 3$
$10 = 9 + 1, 8 = 5 + 3, 6 = 4 + 2$
$10 = 8 + 2, 9 = 6 + 3, 5 = 4 + 1$
$10 = 8 + 2, 9 = 5 + 4, 7 = 6 + 1$
$10 = 8 + 2, 7 = 4 + 3, 6 = 5 + 1$
$10 = 7 + 3, 9 = 8 + 1, 6 = 4 + 2$
$10 = 7 + 3, 5 = 4 + 1, 8 = 6 + 2$
$10 = 6 + 4, 9 = 8 + 1, 7 = 5 + 2$
$10 = 6 + 4, 9 = 7 + 2, 8 = 5 + 3$
$10 = 6 + 4, 8 = 7 + 1, 5 = 3 + 2$

Observe what strategies children use to tackle this question. It is easiest to start with 10 and work backwards (for example, $10 = 9 + 1$, $8 = 6 + 2$ and $7 = 4 + 3$). Support children by giving them whiteboards so that they can try combinations then rub numbers out if they go wrong, by giving them 1–9 digit cards to move about, or by giving them cubes so they can experiment with sharing them out between part-whole models. Children should be prepared to make mistakes and try again until they find a solution.

Power check

WAYS OF WORKING Independent thinking

ASK
- *What did you find the most difficult?*
- *What pictures do you make in your mind to help you answer the questions?*

Power play

WAYS OF WORKING Pair work or small groups

IN FOCUS Use this **Power play** to see if children can work in pairs to complete a part-whole model. Children will need to think carefully about whether to place one or two counters to try to ensure their opponent does not win.

ANSWERS AND COMMENTARY Answers will depend on what number children choose as their whole. Consider pairing up children with varying levels of confidence so that more confident children can support less confident children.

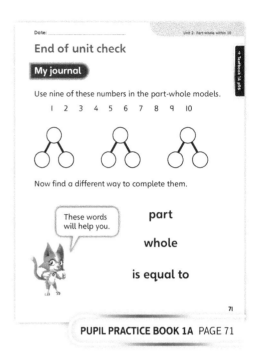

PUPIL PRACTICE BOOK 1A PAGE 71

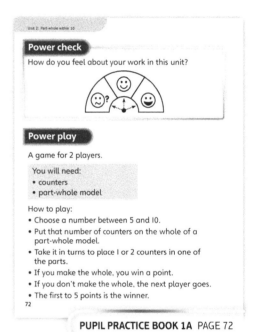

PUPIL PRACTICE BOOK 1A PAGE 72

After the unit ⏸

- How confident were children identifying the parts and the whole and writing number sentences?
- Can children find number bonds without counting on?

Strengthen and **Deepen** activities for this unit can be found in the *Power Maths* online subscription.

Unit 3
Addition within 10

Mastery Expert tip! 'When I taught this unit, I used the characters within the Pupil Books to encourage children to talk about the different methods they were using. This worked well and helped us explore different ways of working. It also encouraged children to talk openly about mistakes.'

Don't forget to watch the Unit 3 video!

WHY THIS UNIT IS IMPORTANT

This unit builds on the work on number bonds within 10 from the previous lessson. It is important that children become fluent in these facts, because they are the foundation for future number facts.

Within this unit, children are introduced to formal addition for the first time through the idea of 'count all' and 'count on' strategies. A 'count all' strategy is when all parts are added together to make a whole. A 'count on' strategy asks children to start with a number and count on.

As well as introducing children to some of the key language associated with addition, children will also begin to develop an understanding of the commutativity of addition – the idea that addition calculations can be performed in any order.

WHERE THIS UNIT FITS

→ Unit 2: Part-whole within 10
→ **Unit 3: Addition within 10**
→ Unit 4: Subtraction within 10

This unit builds on Unit 2: Part-whole within 10, which introduced children to the idea that a whole can be separated into parts of various sizes. Unit 3 focuses on addition within 10.

Before they start this unit, it is expected that children:
• can use the part-whole model to partition a number to 10
• can write and compare number bonds to 10.

ASSESSING MASTERY

Children who have mastered this unit will be able to relate each number in a calculation to what it represents. Children will be able to use a variety of manipulatives to represent addition within 10, including cubes, ten frames, number lines and part-whole models.

Children's confidence in knowing and recognising number facts and number pairs will also start to increase, and children will start to use these to answer simple calculations without manipulatives.

Additionally, children showing mastery would be able to rearrange the order of a calculation to work efficiently, using their knowledge of commutativity.

COMMON MISCONCEPTIONS	STRENGTHENING UNDERSTANDING	GOING DEEPER
Children may not apply number facts and therefore resort to a 'count all' or 'count on' strategy.	Repeat the number fact after counting objects and remind children that they do not need to count each time.	Ask children to solve missing number problems or to create their own number story.
Children may struggle with the transition from a 'count all' to a 'count on' strategy.	Remind children that we start at the first number, but then count on from there. Count each jump together.	Remember to use numbers within 10. It is important to deepen learning rather than moving children on.

UNIT STARTER PAGES

Use these pages to introduce the unit focus to children. You can use the characters to explore different ways of working.

STRUCTURES AND REPRESENTATIONS

Part-whole model: This model helps children understand that two or more parts combine to make a whole. It also helps to strengthen children's understanding of number bonds within 10.

Number line: Number lines help children learn about addition as counting on. They allow children to identify the starting point, the number counted on and the end point.

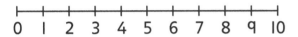

Ten frame: The ten frame helps to give children a sense of 10, and supports their understanding of number bonds to 10. It also plays a key role in helping children to recognise the structure of other numbers, and to understand what happens when you add two numbers together.

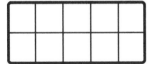

KEY LANGUAGE

There is some key language that children will need to know as part of the learning in this unit.

→ part, whole and part-whole
→ altogether, in total, total, sum
→ add, added, plus, or +
→ count, count on
→ missing, missing part
→ number bonds, number pairs
→ number stories

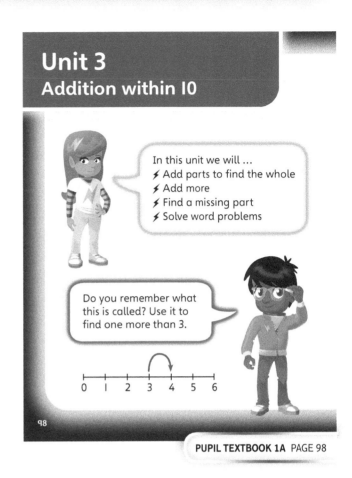

PUPIL TEXTBOOK 1A PAGE 98

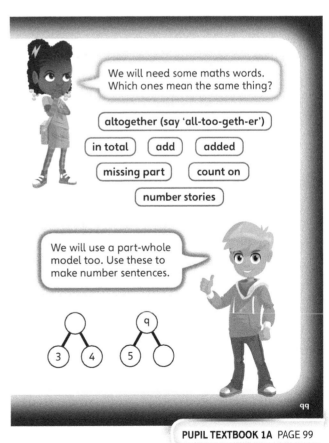

PUPIL TEXTBOOK 1A PAGE 99

Add together

Learning focus

In this lesson, children will combine two parts into a whole and understand how the part-whole model represents addition. Children will make links between concrete representations, part-whole models and abstract addition calculations.

Before you teach

- Based on teaching of the part-whole model in Unit 2, are there any additional misconceptions that need to be addressed?
- How will you support children when moving from structured addition (for example 3 + 4 = ☐) to writing a calculation based on a pictorial representation (for example ☐ + ☐ = ☐)?

NATIONAL CURRICULUM LINKS

Year 1 Number – addition and subtraction

Represent and use number bonds and related subtraction facts within 20.

ASSESSING MASTERY

Children can correctly identify the different parts of an addition calculation and relate them to different pictorial representations or structures. Children can understand that the same whole can be made up of different parts.

COMMON MISCONCEPTIONS

Children may put numbers into the part-whole model incorrectly. They may put the whole as one of the parts, or vice versa. Ask:
- *Which two numbers show the parts? Which number shows the whole?*

Children may also struggle with the idea that the same whole can be made up of different parts. To support understanding, model this with physical resources and ask:
- *The whole is how many cubes there are altogether. If I move the cubes into two different piles and keep changing them, will the whole ever change?*

Children may not link the parts of the addition sentence with the concrete representations. To overcome this use concrete resources such as cubes or counters so that children can create the addition calculations physically and then connect this to the abstract calculation. Ask:
- *What do these counters represent? How do you write this addition in a number sentence? What signs do you use?*

Initially, children may find it confusing that the equals (=) sign appears first in some calculations. Ask:
- *What does the '=' sign mean? Does it mean something different if it comes first?*

STRENGTHENING UNDERSTANDING

You can strengthen understanding by encouraging children to use concrete resources to make their own addition calculations that they can write the addition number sentences for. For each number sentence, help children to split the whole into the two parts correctly.

GOING DEEPER

Encourage children to work systematically to find all the ways to make a whole. You could also expose children to equivalent addition calculations, for example, 4 + 2 = 5 + 1. This is particularly tricky, as there is no defined whole and so the numbers cannot be put straight into a part-whole model.

KEY LANGUAGE

In lesson: add, altogether, '+', added / adding / addition, parts, whole
Other language to be used by the teacher: sum, part-whole model, equal / equals

STRUCTURES AND REPRESENTATIONS

Part-whole model, ten frame

RESOURCES

Mandatory: blank part-whole model, blank ten frames, cubes or counters to put in each of these
Optional: any concrete resources to model addition calculations (for example, cubes, counters, teddies, cars)

 In the eTextbook of this lesson, you will find interactive links to a selection of teaching tools.

Quick recap 🔎

Clap up to ten times to your class. Ask children to count how many times you clap. Repeat, counting aloud together to find the total number of claps.

Discover

Add together

Discover

WAYS OF WORKING Pair work

ASK
- Question ① a): *Which numbers are the parts?*
- Question ① b): *Which number is the whole?*

IN FOCUS In this part of the lesson, we first want children to understand how a real-life context can give numbers that are parts and wholes. The word 'altogether' is used to promote understanding that the two parts can combine to make a whole. They should use their understanding of parts and wholes from the previous unit.

PRACTICAL TIPS Ask children to use counters to represent the children that are standing up and sitting down.

DEEPEN Question ① b) asks children to combine the two parts to make the whole.

ANSWERS

Question ① a): There are 6 children sitting and 4 children standing.

Question ① b): There are 10 children altogether.

① a) How many children are sitting?

How many children are standing?

b) How many children are there altogether?

100

PUPIL TEXTBOOK 1A PAGE 100

Share

WAYS OF WORKING Whole class teacher led

ASK
- Question ① a): *How many children are sitting?*
- Question ① a): *How many children are standing?*
- Question ① b): *How many children are there altogether?*
- Question ① b): *Where do the children sitting go in the number sentence?*
- Question ① b): *Where do the children standing go in the number sentence?*

IN FOCUS In this part of the lesson, the concrete representations are being linked to the abstract calculation. The addition sign, '+', is being introduced in context. Use the characters to introduce this idea.

ASSESSMENT CHECKPOINT In question ① a), assess whether children can point to the two parts (the children sitting and the children standing) and understand how this information relates to a part-whole model.

In question ① b), children should be able to write the information using a part-whole model and a number sentence.

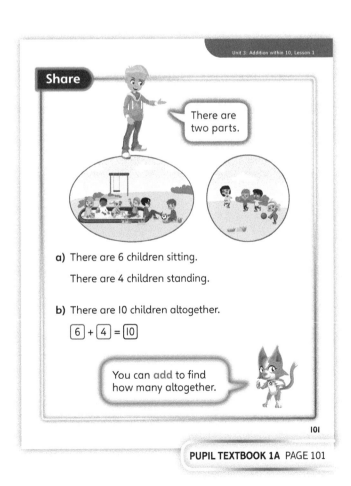

Share

There are two parts.

a) There are 6 children sitting.

There are 4 children standing.

b) There are 10 children altogether.

$$\boxed{6} + \boxed{4} = \boxed{10}$$

You can **add** to find how many altogether.

101

PUPIL TEXTBOOK 1A PAGE 101

Think together

WAYS OF WORKING Whole class teacher led (I do, We do, You do)

ASK

- Question ❶: *How many counters are in the first circle? Where is this number in the number sentence? How many counters are in the second circle? Where is this number in the number sentence? How can we find out how many counters there are altogether?*
- Question ❷: *How many dots are on each dice? How does this help us write the addition? How can we work out the total?*

IN FOCUS In questions ❶ and ❷, children work out simple additions using images where they are still able to count one by one if needed. Children need to understand that one method they can use is to count all of the counters or dots one at a time. Encourage children to take care when counting. Question ❸ moves onto an abstract representation of addition and prompts children to use concrete resources if needed.

STRENGTHEN Encourage children to use counters to represent the images. Ask them to place the counters on a ten frame to find the total.

DEEPEN Encourage children to find the total by counting on from the greatest number rather than from 1. Ask: *Do you need to start with the first number in the addition?* For example, in question ❷, children could count on 3 from 5 to get a total of 8.

ASSESSMENT CHECKPOINT Use question ❸ to check if children can complete the additions and are able to represent them using counters. Children should understand which number in the additions matches what they have made with the counters.

ANSWERS

Question ❶: 3 + 4 = **7**

Question ❷: 5 + 3 = 8

Question ❸ 5 + 2 = **7**
7 + 1 = **8**
2 + 6 = **8**
3 + 3 = **6**

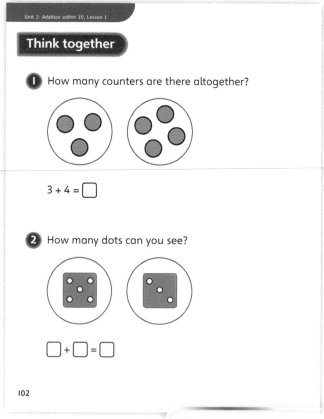

PUPIL TEXTBOOK 1A PAGE 102

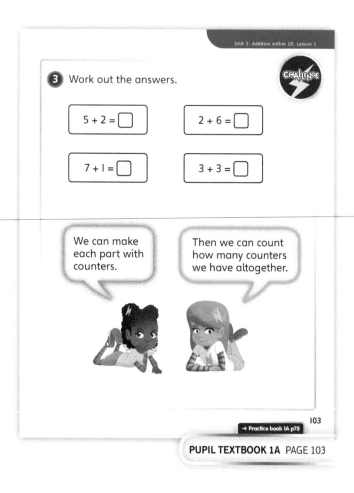

PUPIL TEXTBOOK 1A PAGE 103

Practice

WAYS OF WORKING Independent thinking

IN FOCUS In questions **1**, **2** and **3**, children continue to make links between pictorial and concrete representations of additions. They continue to find the total by counting all the counters, apples or cars. It is not necessary at this stage for children to understand counting on, although some children may prefer to do this as they realise it is more efficient. Question **4** focuses on abstract representations of additions, with children being prompted via Dexter's comment to use concrete objects if needed. Children may already know the bond for some answers and this should be celebrated.

STRENGTHEN Children can use counters or cubes to represent the additions in all of the questions. They can place the counters or cubes on a ten frame to help them find the total.

DEEPEN Encourage children to count on from the greatest number in each addition to find the total. Ask: *Do you need to start with the first number in the addition?*

ASSESSMENT CHECKPOINT Use question **4** to assess whether children are confident with addition. They should be able to make the additions using counters or cubes, find the total and understand what each number in the addition represents.

ANSWERS Answers for the **Practice** part of the lesson can be found in the *Power Maths* online subscription.

Reflect

WAYS OF WORKING Pair work

IN FOCUS In this question, children may count all to find the total or realise that 5 + 5 = 10 through their subitising skills.

ASSESSMENT CHECKPOINT Check if children recognise that 5 is half of a ten frame, and thus realise that 5 + 5 must be equal to 10. It is fine, however, if children do not recognise this and count on as they have done previously.

ANSWERS Answers for the **Reflect** part of the lesson can be found in the *Power Maths* online subscription.

After the lesson ⏸

- Are children confident with what each of the numbers in each structure represents, and can they move between them and the concrete examples?

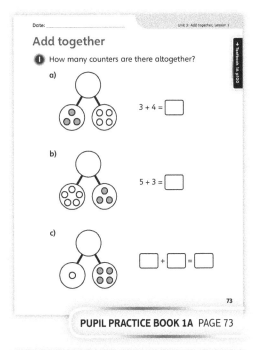

PUPIL PRACTICE BOOK 1A PAGE 73

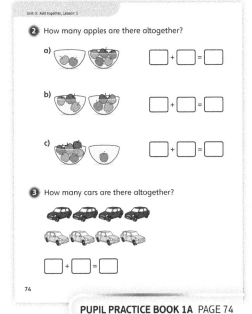

PUPIL PRACTICE BOOK 1A PAGE 74

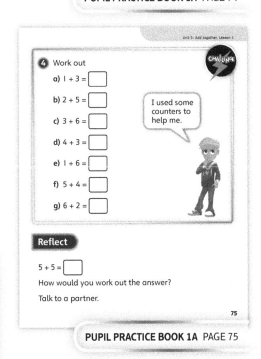

PUPIL PRACTICE BOOK 1A PAGE 75

Add more

Learning focus

In this lesson, children will be able to find a total by counting on from one amount rather than having to start at 0.

Before you teach

- Based on teaching of the part-whole model and '+' sign in Lesson 1, are there any remaining misconceptions that need to be addressed?
- How will you explore common mistakes, such as counting on from the starting number instead of the next number? Can children be encouraged to circle their starting number on a number line before counting on?

NATIONAL CURRICULUM LINKS

Year 1 Number – addition and subtraction

Represent and use number bonds and related subtraction facts within 20.

ASSESSING MASTERY

Children can accurately count on from an amount, and represent this on a number line, showing the amount being added on as a number of jumps. Children can also develop their understanding of the commutative law of addition: for example, $2 + 7 = 7 + 2$.

COMMON MISCONCEPTIONS

Children may incorrectly use their starting number as their first count. When calculating the answer to $5 + 2$, for example, children may count the number 5 as their first count. They need to recognise that each count is in a one-to-one relationship with the objects being added. To support understanding, model this with concrete resources and ask:

- *If I have 5 cubes to start with, and add 2 more, what number is my first count?*

Children may also lose track of how many they are counting on, as they have to hold the number they are counting on in their head as well as counting out loud from another number. One way of overcoming this is to ask:

- *Can you hold up fingers or use cubes to match the number that you are counting on?*

STRENGTHENING UNDERSTANDING

To strengthen understanding, you can encourage children to count on from any given number to 10. Pairs could take it in turns to pick a number less than 10 and ask their partner to count on 5 / count up to 10.

GOING DEEPER

You can encourage children to deepen understanding by building on prior learning of 'more than' or 'less than'. For example, model expressing $5 + 2 = 7$ as 2 *more than* 5 is 7.

KEY LANGUAGE

In lesson: count on, add / added / adding / addition, plus, '+', in total

Other language to be used by the teacher: add more, starting point, total, altogether, jumps

STRUCTURES AND REPRESENTATIONS

Ten frame, number line

RESOURCES

Mandatory: digit cards, blank ten frames, number lines, marbles/cubes, a jar

Optional: number tracks, bead strings, multilink cubes

 In the eTextbook of this lesson, you will find interactive links to a selection of teaching tools.

Quick recap ♫

Make some digit cards from 0 to 9. Ask a child to choose a card, then ask them to count on to 10 from this number.

Discover

WAYS OF WORKING Pair work

ASK

• Question **1** a): *How many marbles are already in the jar? How many marbles are in the teacher's hand? Can you show this with counters and a ten frame? How many marbles are there altogether?*

IN FOCUS In question **1** a), children are presented with a real-life context for 'adding on' a certain amount when starting from an amount above 0. 'First, then, now' stories are a useful way of framing an addition problem.

PRACTICAL TIPS Use counters and ten frames to represent the story.

ANSWERS

Question **1** a): First there were 5 marbles.
Then the teacher added 2 marbles.
Now there are 7 marbles.

Question **1** b): 5 + 2 = 7

Add more

Discover

I am adding 2 marbles to the jar.

There are 5 marbles in the jar.

1 a) Complete the story.

First …

Then …

Now …

b) Write a number sentence for your story.

104

PUPIL TEXTBOOK 1A PAGE 104

Share

WAYS OF WORKING Whole class teacher led

ASK

• Question **1** a): *What do the 5 and 2 refer to? Why did we count on from 5? Why did we count two jumps? Why did we count on … 6 … 7 … instead of … 5 … 6 …?*

IN FOCUS This part of the lesson lends itself to being very practical and sensory, as children can learn to count on using sounds. Use concrete resources such as marbles or cubes to model the story in real life. Tell children there are 5 marbles (or cubes) to begin with, and then ask them to count on with every sound they hear (as each cube/marble is dropped into a jar). This reinforces one-to-one correspondence. Alternatively, you could ask them to count on when you show them one marble (cube) at a time.

Share

a)

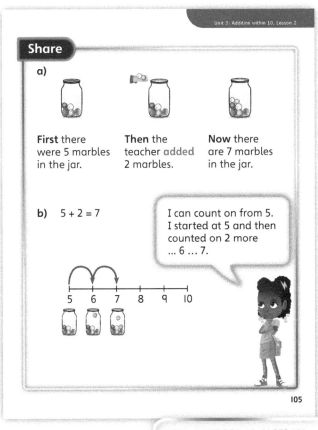

First there were 5 marbles in the jar.

Then the teacher **added** 2 marbles.

Now there are 7 marbles in the jar.

b) 5 + 2 = 7

I can count on from 5. I started at 5 and then counted on 2 more … 6 … 7.

105

PUPIL TEXTBOOK 1A PAGES 105

143

Think together

WAYS OF WORKING Whole class teacher led (I do, We do, You do)

ASK

- Question **1** a): *What is our starting point? How many marbles were in the jar to start with? Where do we see this number on the number line?*
- Question **1** a): *How many marbles are being added? Where is that shown on the number line?*
- Question **2**: *Where is our starting number on the number line? How many are being added? How is that shown on the number line? What is the total?*
- Question **3**: *Does it make most sense to start counting from the smallest or the largest number in this calculation?*

IN FOCUS Question **3** encourages children to think carefully about the most efficient strategy, and to decide from which number to count on: the smallest or the largest. Refer to the characters discussing the different start numbers. Ask: *Can you point to the starting point on the number line?*

STRENGTHEN For question **3**, you could use multilink cubes or bead strings to reinforce the practice of counting on. Children can represent the calculation by physically moving the number of multilink cubes or beads they need to add on.

DEEPEN This part of the lesson provides a good opportunity to ensure that children can check their answers by counting backwards.

ASSESSMENT CHECKPOINT Assess whether children are able to relate the parts of a calculation to the number line accurately. For example, in question **2**, 5 is the starting point and 3 is added. Therefore, 5 should be the starting point on the number line and there should be three jumps.

ANSWERS

Question **1** a): First there were 6 marbles.
Then 2 marbles were added.
Now there are 8 marbles.

Question **1** b): $6 + 2 = 8$

Question **2**: $5 + 3 = 8$

Question **3**: $2 + 7 = \mathbf{9}$

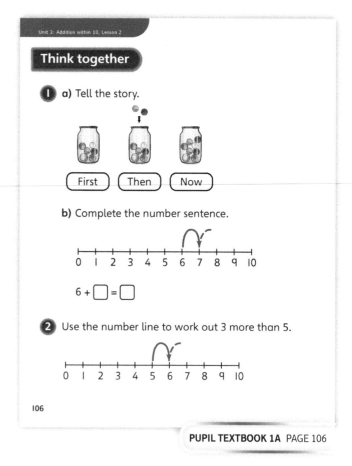

PUPIL TEXTBOOK 1A PAGE 106

PUPIL TEXTBOOK 1A PAGE 107

Practice

WAYS OF WORKING Independent thinking

IN FOCUS In questions ① a) and b), the starting point has been given to children in the calculation. In question ① c), children need to find the starting point for themselves, and then count on. Question ② looks at examples of counting on from a given number. Children use a number line to show the jumps. They realise where they finish their jumps gives the answer to the addition. Children should start to understand that it does not matter which number they start with, they will always get the same answer. Children may therefore start to see that they could always start with the greatest number. Some children may know the answer to the additions from knowledge of bonds and this should be celebrated. They can use the number line to prove the answers.

STRENGTHEN In all questions, help children to realise that they do not need to count the start number, but simply find it on the number line. In question ① a), reinforce the fact that the number of fingers on one hand is 5 by asking: *Can you show me 5 on your hands? How can you show me adding on 2?* Alternatively, children could show this using cubes. In question ②, children could use number lines and counters to help them.

DEEPEN Some children may be able to make one jump on a number line rather than several jumps.

ASSESSMENT CHECKPOINT Check that children are starting on the correct point on the number line, and counting on the correct amounts. Children should not be starting at 0 to work out the overall total, even if their answers are correct.

ANSWERS Answers for the **Practice** part of the lesson can be found in the *Power Maths* online subscription.

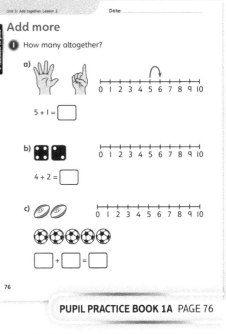

PUPIL PRACTICE BOOK 1A PAGE 76

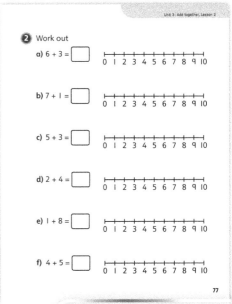

PUPIL PRACTICE BOOK 1A PAGE 77

Reflect

WAYS OF WORKING Pair work, whole class

IN FOCUS Children are asked to explain how to work out the total of two numbers. They should not just give the answer, but demonstrate how they worked it out.

ASSESSMENT CHECKPOINT Assess whether children use words, a model or a number line to explain their method. Does their method give the correct total?

ANSWERS Answers for the **Reflect** part of the lesson can be found in the *Power Maths* online subscription.

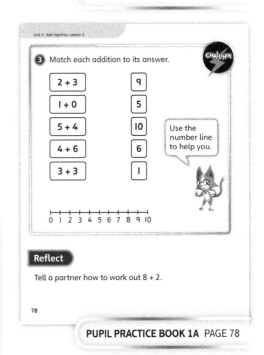

PUPIL PRACTICE BOOK 1A PAGE 78

After the lesson

- Have children mastered counting on, rather than counting from 0, to find a total?
- Did children recognise when their approach was correct but took longer, for example, by starting with the smaller part and counting on the larger?

Addition problems

Learning focus

In this lesson, children will find solutions to simple word and picture problems involving additions to 10.

Before you teach

- Are children confident with the fundamental concept that two parts join together to make a whole, and recognise it on a part-whole model and in an addition calculation? If not, the real life contexts in this lesson may help children.
- What links could you make with other curriculum areas? Could curriculum links help children to compose their own word problems?

NATIONAL CURRICULUM LINKS

Year 1 Number – addition and subtraction

Solve one-step problems that involve addition and subtraction, using concrete objects and pictorial representations, and missing number problems such as $7 = \boxed{} - 9$.

ASSESSING MASTERY

Children can find number stories in pictures and use addition to answer questions. Children can represent contexts using addition calculations and explain the meanings of the parts of their calculations.

COMMON MISCONCEPTIONS

The number of objects in the pictures are not always represented in an ordered way, which may encourage children to go back to counting from 0 to find the total. Draw links from previous lessons and model that once you have found how many objects are in one group, you can use that number to count on. Ask:
- *How many groups can you see?*

STRENGTHENING UNDERSTANDING

The contexts explored in this lesson show that number stories and number sentences describe the real world, in a similar way to that in which words and pictures do. Strengthen understanding by encouraging children to come up with their own context for one of the calculations. For example, instead of 6 jam tarts + 3 jam tarts, they could use 6 bears + 3 bears.

GOING DEEPER

Encourage children to explore this lesson in more depth by asking them to make up their own word problem, and to draw a picture that represents it, with the matching calculation alongside.

KEY LANGUAGE

In lesson: number stories, altogether, count, in total

Other language to be used by the teacher: parts, whole, addition calculation, greater than

STRUCTURES AND REPRESENTATIONS

Part-whole model, number line

RESOURCES

Mandatory: blank part-whole models, number lines
Optional: classroom objects / PE equipment

 In the eTextbook of this lesson, you will find interactive links to a selection of teaching tools.

Quick recap

Play addition bingo. Ask children to pick four numbers between 0 and 10. Write some addition calculations that have answers between 0 and 10. If children have the answer, they cross it off.

Discover

Addition problems

WAYS OF WORKING Pair work

ASK

• Question ❶ a): *What do the 4 and the 4 mean?*
• Question ❶ b): *Where on the picture can you see 1 person with 3 other people?*

IN FOCUS The picture contains lots of hidden number stories. Explain that you will be exploring them together throughout this lesson. Question ❶ a) focuses attention on one part of the picture (people walking dogs) and elicits a number story from children. Question ❶ b) focuses on a different part and a different story (the seesaw).

PRACTICAL TIPS Use concrete objects in the classroom to represent the scenario.

ANSWERS

Question ❶ a): There are 8 dogs altogether.

Question ❶ b): There is 1 person on one side of the seesaw, and 3 people on the other side: $1 + 3 = 4$.

Discover

Dale   Liz

❶ a) Dale has 4 dogs and Liz has 4 dogs.

   How many dogs are there altogether?

b) What part of the picture shows $1 + 3$?

   Show this using a part-whole model.

108

PUPIL TEXTBOOK 1A PAGE 108

Share

WAYS OF WORKING Whole class teacher led

ASK

• Question ❶ a): *What does the number 4 show in this calculation?*
• Question ❶ b): *What different thing does the number 4 show in this calculation?*
• Question ❶ b): *If you added 1 dog and 3 dogs, would that be the same as adding 1 ball and 3 balls?*

IN FOCUS Before question ❶ a), two different characters suggest different methods for solving the calculation. Look at what Astrid says. Ask: *Do you agree or disagree with Astrid? Counting using counters would give us the right answer, but it would also take a long time.*

STRENGTHEN Strengthen children's understanding by discussing that the 4 in question ❶ a) represents the number of dogs one person has (a part). The 4 in question ❶ b) represents the number of people altogether (the whole). Ask: *Can you spot any other places in the picture where one number has different meanings?*

Share

I used counters.

I used a part-whole model and tried a number line too.

a)

8

4   4

$4 + 4 = 8$

0 1 2 3 4 5 6 7 8 9 10

There are 8 dogs altogether.

b)

4

1   3

$1 + 3 = 4$

There are 4 people on the see-saw.

109

PUPIL TEXTBOOK 1A PAGE 109

147

Think together

WAYS OF WORKING Whole class teacher led (I do, We do, You do)

ASK

- Question ❶: *How many jam tarts can you see? How many jam tarts are hidden?*
- Question ❷: *You can see the whole picture and all of the parts of the story. How many children have short hair? How many children have long hair?*

IN FOCUS In question ❶, there are 3 jam tarts shown and 6 hidden in the box. Children may be tempted to start counting on from 3 on the number line, as that is the number they can see. In question ❷, children develop first their confidence in knowing the parts and how they can use different methods to find the whole. For example, some children may just count all the children on the roundabout, others may count on. Children realise that an addition can be written down either way as the answer is the same. So 2 + 3 gives the same answer as 3 + 2. Children could use concrete materials to show that this is the case.

STRENGTHEN Strengthen children's understanding by asking them to work in pairs to complete question ❸, which prompts them to look again at the picture from the **Discover** section. Ask children to tell their partners another number story they can see. Their partner should articulate which number relates to which part of the story.

DEEPEN Deepen children's understanding of question ❶ by asking them to switch the numbers around. In the question, there are 6 tarts in the box and 3 tarts on the plate.

ASSESSMENT CHECKPOINT Assess how far children understand that calculations have meanings based on the contexts of the problems. Check that they grasp that changing contexts does not impact the answer.

ANSWERS

Question ❶: 6 + 3 = 9
There are 9 jam tarts altogether.

Question ❷: 2 + 3 = 5
3 + 2 = 5

Question ❸: For example:
There are 2 adults and 4 children at the picnic.
2 + 4 = 6.
There are 6 people at the picnic altogether.

Think together

❶ There are 6 jam tarts in the box.

There are 3 jam tarts on the plate.

How many jam tarts are there altogether?

6 Jam Tarts

⬜ + ⬜ = ⬜

110

PUPIL TEXTBOOK 1A PAGE 110

---

❷ Look at the children on the roundabout.

There are 2 children with long hair.

There are 3 children with short hair.

There are 5 children altogether.

Find two ways to show this fact.

⬜ + ⬜ = ⬜   ⬜ + ⬜ = ⬜

❸ Look at the park in Discover.
What other **number stories** can you see?
Ask a partner.
How many number stories can you find?

CHALLENGE

I will try to find a number story that no one else has seen.

111

→ Practice book 1A p79

PUPIL TEXTBOOK 1A PAGE 111

Practice

WAYS OF WORKING Independent thinking

IN FOCUS Children interpret each image to create number sentences. In questions ① and ②, children could count to find the total. Questions ③ onwards removes this scaffold.

STRENGTHEN If needed for questions ③ onwards, children could use counters or cubes to represent the numbers, giving them the option to count to find the total. If children do this, encourage them to start counting from the greater number rather than from 1.

DEEPEN In question ⑤, challenge children to draw a picture where it is possible to count and a picture where it is not possible to count.

ASSESSMENT CHECKPOINT Check that children are counting on to find the totals. Identify children who need to use concrete materials to access the questions.

ANSWERS Answers for the **Practice** part of the lesson can be found in the *Power Maths* online subscription.

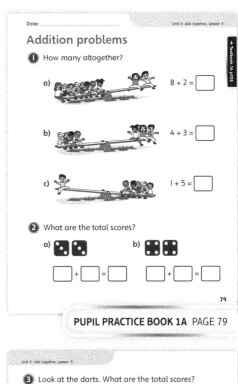

PUPIL PRACTICE BOOK 1A PAGE 79

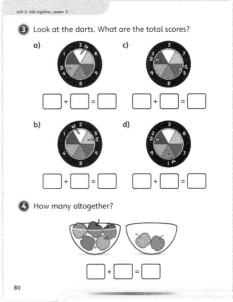

PUPIL PRACTICE BOOK 1A PAGE 80

Reflect

WAYS OF WORKING Group work

IN FOCUS Children work backwards to tell a story for 4 + 1. They may use a 'first, then, now' story to do this.

ASSESSMENT CHECKPOINT Listen for children's language in explaining their story. What do they say? Do they use the words and phrases 'plus', 'is equal to', 'altogether' and 'in total'?

ANSWERS Answers for the **Reflect** part of the lesson can be found in the *Power Maths* online subscription.

After the lesson ⏸

- Are children fluent enough with their number bonds to 10 that they are using them in a variety of contexts, including contexts they have made up?
- Have any children shown an awareness of counting back, as well as counting on, when using the number line? This will be explored further in the next unit.

PUPIL PRACTICE BOOK 1A PAGE 81

Find the missing number

Learning focus

In this lesson, children will use what they have learnt about addition to solve missing number problems. It is important that children think about addition and parts and wholes to help them solve these problems rather than be formally introduced to subtraction at this stage.

Before you teach

- Are children confident identifying the parts and wholes in a part-whole model? Spend time reinforcing this or children may find it difficult to access this lesson.
- Do children understand what 'adding more' means?
- Are children confident starting at any number when counting on or are they still making mistakes? Using counters and a ten frame will help them.

NATIONAL CURRICULUM LINKS

Year 1 Number – number and addition

Represent and use number bonds and related subtraction facts within 20.

ASSESSING MASTERY

Children can use their knowledge of addition and parts and wholes to solve missing number problems. Children count on from any number rather than starting at 0 every time.

COMMON MISCONCEPTIONS

In this lesson, children will be finding a missing part. They may mistake finding a part for finding the whole and simply add together the numbers they can see. Ask:
- *What does each number represent?*

STRENGTHENING UNDERSTANDING

The use of the part-whole model is important in this lesson. Spend time simply identifying where the whole is and where the parts are in part-whole models.

GOING DEEPER

By now some children may be becoming fluent with their number bonds. Do not discourage their instant recognition of the missing part but ensure they are able to explain and show how they know.

KEY LANGUAGE

In lesson: part, whole, **missing**, **part**, add, more

Other language to be used by the teacher: show, represent

STRUCTURES AND REPRESENTATIONS

Part-whole model, ten frame

RESOURCES

Mandatory: part-whole model, counters/cubes

Optional: ten frames, toys

 In the eTextbook of this lesson, you will find interactive links to a selection of teaching tools.

Quick recap 𝛺

Give children between five and ten double sided counters. Ask them to find all the bonds to a given number using the counters. How many did they know off by heart?

Discover

Unit 3: Addition within 10, Lesson 4

Find the missing number

Discover

WAYS OF WORKING Pair work

ASK

- Question ① a): *How many elephants are on the shelf? How many more elephants will make 6 altogether?*

IN FOCUS In questions ① a) and b), children consider a real-life context to begin to explore finding missing parts.

PRACTICAL TIPS Bring the scene to life using some real toys from school, for example, four teddy bears or four toy cars. Add more together as a class and count until you make 6.

ANSWERS

Question ① a): $4 + \mathbf{2} = 6$

Question ① b):

① a) How many more elephants to make 6?

$4 + \boxed{} = 6$

b) Show this with a part-whole model.

112

PUPIL TEXTBOOK 1A PAGE 112

Share

WAYS OF WORKING Whole class teacher led

ASK

- Question ① a): *How many more elephants would we add to the shelf to make 6 elephants altogether? How would we write a number sentence for this?*
- Question ① b): *Look at your number sentence. Which numbers are the parts? Which number is the whole?*

IN FOCUS In question ① a), children use a real-life example to exemplify a missing number problem. They use the method of counting on or their knowledge of number bonds to find the missing number. In question ① b), children need to be able to identify what the elephants represent in the number sentence and in the part-whole model.

Share

a) Start with 4 elephants.
Add 2 to make 6.

$4 + ② = 6$

I counted on to find the missing part.

b)

6 is the whole. 4 is a part.
The part that was missing is 2.
$4 + 2 = 6$

113

PUPIL TEXTBOOK 1A PAGE 113

Think together

WAYS OF WORKING Whole class teacher led (I do, We do, You do)

ASK

- Question **1**: *How many more counters do we need to add to make 5? Should we start counting at 1 or 3? Why? What links can you see between the ten frame and the number sentence?*
- Question **2**: *Are the blank circles parts or wholes? Will the missing number be bigger or smaller than the whole?*

IN FOCUS Children find missing numbers using their knowledge of parts and wholes. It is important that children link what they make using counters and a ten frame to the part-whole model and the number sentences.

STRENGTHEN Use two different colours of counters to emphasise the different parts.

DEEPEN Some children may be ready to try the challenge questions without the use of any concrete manipulatives and part-whole models. Ask: *What can you see in your head when you answer the questions?*

ASSESSMENT CHECKPOINT Children can find missing parts independently and may have started to become fluent in their number bonds within 10.

ANSWERS

Question **1** a): Children should add two counters to the ten frame.

Question **1** b): $3 + \mathbf{2} = 5$

Question **2** a): 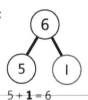 b):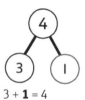

$3 + \mathbf{1} = 4$ $\mathbf{2} + 2 = 4$

Question **3**: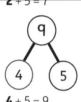

$5 + \mathbf{1} = 6$ $\mathbf{2} + 5 = 7$

$5 + \mathbf{3} = 8$ $\mathbf{4} + 5 = 9$

Think together

1 a) Add more counters to make 5.

b) How many more did you add?

$3 + \boxed{} = 5$

2 Find the missing parts.

a) 4 / 3 / ◯

$3 + \boxed{} = 4$

b) 4 / ◯ / 2

$\boxed{} + 2 = 4$

I will use counters on a ten frame.

114

PUPIL TEXTBOOK 1A PAGE 114

3 Find the missing parts.

CHALLENGE

6 / 5 / ◯

$5 + \boxed{} = 6$

7 / ◯ / 5

$\boxed{} + 5 = 7$

8 / 5 / ◯

$5 + \boxed{} = 8$

9 / ◯ / 5

$\boxed{} + 5 = 9$

Is there a pattern?

I will count on to find the answer.

115

→ Practice book 1A p82

PUPIL TEXTBOOK 1A PAGE 115

Practice

WAYS OF WORKING Independent thinking

IN FOCUS Children answer missing number problems in a variety of ways. In question ❶, they begin by drawing more objects to find the total. In question ❷, children complete a ten frame to help them find the missing number. In question ❸, children use the number line to consider how many they need to count on to find the missing the number. Children progress to more abstract calculations as they go.

STRENGTHEN When children progress to the more abstract questions, encourage them to use the drawing techniques they used in the earlier questions or use counters and a ten frame to represent the additions.

DEEPEN Encourage children to spot patterns within number sentences. You could create some questions similar to question ❺. For example, 5 + ☐ = 8 and 5 + ☐ = 7. Ask: *What do you notice about the missing numbers? Why is this the case?*

ASSESSMENT CHECKPOINT Questions ❹ and ❺ assess whether children have grasped the concept of missing number problems with addition. It is not essential that all children can notice the links in question ❺ but it is a good discussion to have as a class.

ANSWERS Answers for the **Practice** part of the lesson can be found in the *Power Maths* online subscription.

Reflect

WAYS OF WORKING Whole class

IN FOCUS Children explain how to find a missing number in an addition. Encourage them to show this using counters and a ten frame, alongside a part-whole model. Some children may even be able to show others what a common mistake might be (adding together the two numbers seen in the question).

ASSESSMENT CHECKPOINT Children can explain how to find a missing number and make up their own problem for someone else to solve.

ANSWERS Answers for the **Reflect** part of the lesson can be found in the *Power Maths* online subscription.

After the lesson

- When given a real-life context, can children identify the parts and whole?
- Can children confidently identify and explain what the parts and the whole are in number sentences and part-whole models?
- Can children use their knowledge of parts and wholes to solve missing number problems?

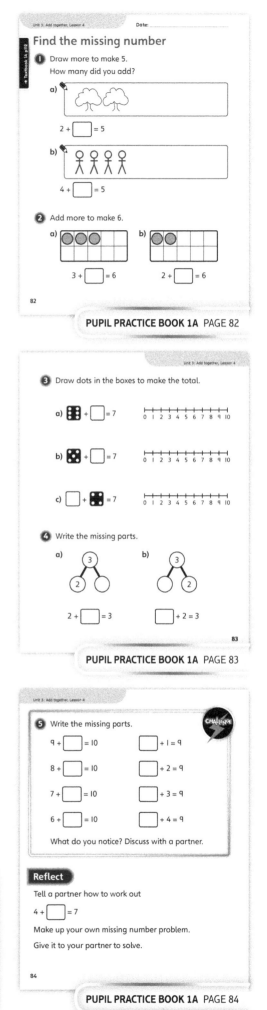

PUPIL PRACTICE BOOK 1A PAGE 82

PUPIL PRACTICE BOOK 1A PAGE 83

PUPIL PRACTICE BOOK 1A PAGE 84

End of unit check

> Don't forget the unit assessment grid in your *Power Maths* online subscription.

WAYS OF WORKING Group work teacher led

IN FOCUS Question ❸ simply requires children to complete a number sentence. It will allow you to identify those children who can work the answer out mentally, and those who still need equipment or other support.

Think!

WAYS OF WORKING Pair work or small groups

IN FOCUS This question involves a lot of deep thinking. Any of the three choices could be the odd one out, as long as children's explanations are mathematically sound and appropriate (for instance, saying that two choices use the number 5, but the other does not, would not be acceptable).

When listening to explanations, encourage children to use the key vocabulary shown at the bottom of the **My journal** page.

Encourage children to think through or discuss the possible answers before writing their answer in **My journal**.

ANSWERS AND COMMENTARY Children who have mastered the concepts of this unit will be able to add and subtract using efficient methods, knowing when and how to use equipment to help them. They will be secure in their use of number bonds to 10 and be able to find different ways to make the same total. They will use the part-whole model and number line with confidence and be able to solve word problems involving addition or subtraction.

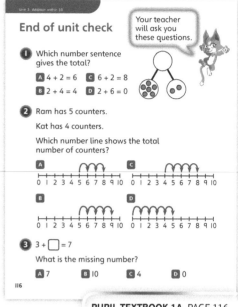

PUPIL TEXTBOOK 1A PAGE 116

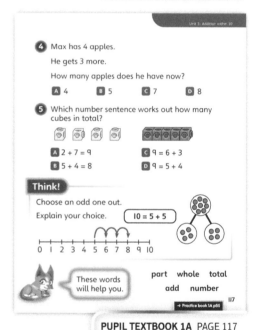

PUPIL TEXTBOOK 1A PAGE 117

Q	A	WRONG ANSWERS AND MISCONCEPTIONS	STRENGTHENING UNDERSTANDING
1	C	A suggests that children have not fully understood the part-whole model, thinking incorrectly that the first part (which is 6) is actually the whole.	Give children further daily counting support. Short, sharp sessions in which counters and number lines are used are very effective.
2	B	D suggests that children are not confident with counting on from a number greater than 0. They have counted up to 5 but have not then counted on 4 from 5.	Revise the structure of the part-whole model using apparatus: hoops and counters reinforce it well. A secure understanding of this model is important as a foundation for more complex work later on.
3	C	B suggests that children are confused by missing numbers and do not understand the function of the equals sign.	
4	C	Not choosing C suggests that children are not confident counting on from the correct start number.	Give children number sentences to rearrange. For example, give them $2 + 3 = 5$ and show them they could also write it as $3 + 2 = 5$, $5 = 3 + 2$, $5 = 2 + 3$.
5	D	Not choosing D suggests that children are unfamiliar with number sentences in which the answer appears first.	

My journal

WAYS OF WORKING Independent thinking

ANSWERS AND COMMENTARY

Any of the three options could be the odd one out.

Possible explanations include:
- 10 = 5 + 5: Because 5 + 3 = 8 and 4 + 4 = 8 are both ways of partitioning 8.
- 10 = 5 + 5: Because the number line and part-whole model are both representations, not number sentences.
- 5 + 3 = 8: Because 4 + 4 = 8 and 5 + 5 = 10 both show doubling. Although children will not know the term 'doubling', they may be able to identify and explain the pattern of doubling in these calculations.

Assess whether children use the correct vocabulary in their answers. Model exemplary answers to children. For example, if a child says, *I choose 5 + 5 = 10 as the odd one out, because the other two both equal 8,* say back to them, *Yes, the whole is 8 for both of them and the two parts make a total of 8, but in different ways.*

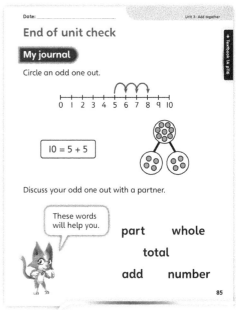

PUPIL PRACTICE BOOK 1A PAGE 85

Power check

WAYS OF WORKING Independent thinking

ASK
- *Do you think you could teach a partner something you have learnt in this unit?*
- *Why did you choose the puzzled face?*

Power play

WAYS OF WORKING Pair work or small groups

IN FOCUS Use the **Power play** to see if children can add pairs of 1-digit numbers mentally.

ANSWERS AND COMMENTARY If children can do the **Power play**, they are secure with 1-digit addition strategies; watch to see if children need to use their fingers or equipment, or make markings on paper. If they cannot play the game, establish whether the problem is with basic counting or with the addition itself. Intervention activities to strengthen children's understanding will be necessary: counting games or adding games would work well; the use of equipment and structures such as the number line is imperative.

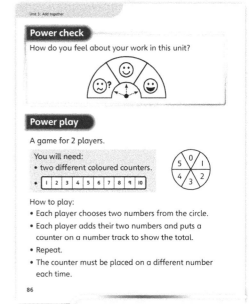

PUPIL PRACTICE BOOK 1A PAGE 86

After the unit ⏸

- What were children's strengths and weaknesses during the unit?
- How many children mastered the unit and how many children need intervention?

Strengthen and **Deepen** activities for this unit can be found in the *Power Maths* online subscription.

Unit 4
Subtraction within 10

Mastery Expert tip! 'Cubes or counters work well when taking away and counting what is left, and a number line works well when counting back. Include lots of chanting on and back, including skip counting back, so children get used to hearing numbers going down as well as up.'

Don't forget to watch the Unit 4 video!

WHY THIS UNIT IS IMPORTANT

This unit focuses on subtraction within 10. Children will be introduced to the key language of subtraction and a range of scenarios in which subtraction takes place.

Within this unit, children are introduced to formal subtraction for the first time: they count how many are left, and break apart a whole. Children will model each of these situations using concrete and pictorial representations: taking away cubes, crossing out pictures and counting back on a number line.

This unit builds on children's previous knowledge of number bonds to 10 and makes explicit links between subtraction facts and previously learnt addition facts. Children are asked to reason with subtraction facts and compare them to numbers using the < and > signs.

WHERE THIS UNIT FITS

→ Unit 3: Addition within 10

→ **Unit 4: Subtraction within 10**

→ Unit 5: 2D and 3D shapes

This unit builds children's knowledge of number bonds within 10, their ability to use a number line to count on and back, and their understanding that two parts make a whole, in the context of subtraction. It looks at subtraction as the inverse of addition and teaches children to count back and work out a missing part, given the whole and the other part. Unit 5 will focus on the properties of 2D and 3D shapes.

Before they start this unit, it is expected that children:

- know how to count back from any number less than 10
- understand the different components of a part-whole model and what each component represents
- know what the < and > signs mean.

ASSESSING MASTERY

Children who have mastered this unit will be able to correctly identify the parts and whole in subtraction calculations even when the = sign is in different places. They will understand that subtraction, unlike addition, is not commutative. Children should be confident in using the part-whole model to represent subtraction and see 'numbers within the numbers'. When there is a subtraction fact where the two parts are the same, children will be able to explain how they still represent different things in context.

COMMON MISCONCEPTIONS	STRENGTHENING UNDERSTANDING	GOING DEEPER
Children may not understand that subtraction is not commutative and may switch numbers around in a subtraction number sentence.	Ask children to model subtraction number sentences using physical resources and explain what each part represents.	Show children a part-whole model and ask them to come up with as many different addition and subtraction number sentences as they can.
Children may incorrectly count back when using a number line so that the starting number is counted as one jump.	When jumping back on a number line, ask children to physically jump back and count each time they do it.	Give children a series of addition and subtraction number sentences and ask them to compare them using < and > signs.

UNIT STARTER PAGES

Use these pages to introduce the unit focus to children. You can use the characters to remind children of the part-whole model and the language used for subtraction.

STRUCTURES AND REPRESENTATIONS

Part-whole model: This model helps children understand that two parts combine together to make a whole and that if you take one part away from the whole, you are left with the other part.

Ten frame: The ten frame gives children a structure in which to put their starting number of objects. They can then remove or cross out the number being taken away and count what is left.

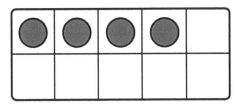

Number line: Number lines help children count back and keep track of how many they are counting back. They allow children to identify the starting point, the number counted back and the end point.

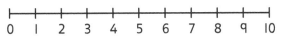

KEY LANGUAGE

There is some key language that children will need to know as part of the learning in this unit.

→ how many are left?

→ take away, subtract

→ subtraction, addition

→ count back

→ more than, >, less than, <

→ missing part

→ number stories

→ fact family

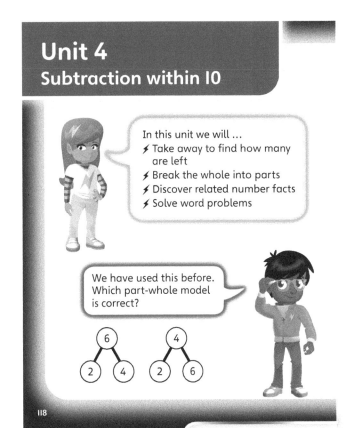

PUPIL TEXTBOOK 1A PAGE 118

PUPIL TEXTBOOK 1A PAGE 119

How many are left?

Learning focus

In this lesson, children will work out simple 'how many are left?' subtractions within 10 by crossing out.

Before you teach

- Are children secure with the contexts provided for subtraction?
- What physical resources could you provide to assist children in making the problems concrete?

NATIONAL CURRICULUM LINKS

Year 1 Number – addition and subtraction

Represent and use number bonds and related subtraction facts within 20.

ASSESSING MASTERY

Children can solve subtractions within 10 by crossing out or physically removing objects and counting how many are left. Children can use contexts to explain their answers and differentiate between the total number to begin with, the number taken away and the number left.

COMMON MISCONCEPTIONS

Children may get confused when using a pictorial representation with crossed out objects, as the total number of objects remains the same. As per the lesson title, it is best to emphasize the word 'are' in the question 'How many *are* left?' Ask:
- *What does it mean if something is crossed out?*

Children may interpret the vocabulary 'How many are left?' as 'How many have left and gone away', rather than 'How many remain, after some are taken away'. Ask:
- *Where does it show you how many are left behind?*

STRENGTHENING UNDERSTANDING

Strengthen understanding by using uninflated balloons, pictures of balloons or counters/cubes to model the concrete situation, and make memorable links between it and different representations. For example, children could hold onto inflated (but not tied) balloons in a line, and then release one so it whooshes around the room.

GOING DEEPER

Discuss with children that, if you remove two objects, you can count what remains or you can count back twice from the total amount. Encourage pairs to deepen understanding by using both methods for some given calculations.

KEY LANGUAGE

In lesson: how many are left?

Other language to be used by the teacher: take away, remain, in total, begin with

STRUCTURES AND REPRESENTATIONS

Number line

RESOURCES

Mandatory: cubes, counters, digit cards

Optional: balloons/pictures of balloons

In the eTextbook of this lesson, you will find interactive links to a selection of teaching tools.

Quick recap 🔄

Make some digit cards from 1 to 10. Ask children to choose a digit card and then count back from this number to 0.

Discover

Unit 4: Subtraction within 10, Lesson 1

How many are left? ❶

WAYS OF WORKING Pair work

Discover

ASK

- Question ❶ a): *How many balloons does Fern have at the start? What happened to one of them? Is Fern still holding it?*
- Question ❶ a): *How many balloons are floating away? How many are left?*
- Question ❶ a): *Why do we not need to count the balloon that is floating away?*
- Question ❶ a): *How could you work this out if there were no pictures?*

❶ a) Fern has 6 balloons.

I balloon floats away.

How many balloons are left?

b) Now I more balloon floats away.

How many balloons are left now?

IN FOCUS Question ❶ a) brings attention to the balloon floating away. The context should help children understand that the balloon is no longer there, and will no longer be counted when they are asked how many are left.

PRACTICAL TIPS Use uninflated balloons, pictures of balloons or counters or cubes to model the scenario.

ANSWERS

Question ❶ a): There are 5 balloons left.

Question ❶ b): There are 4 balloons left.

120

PUPIL TEXTBOOK 1A PAGE 120

Share

Unit 4: Subtraction within 10, Lesson 1

Share

WAYS OF WORKING Whole class teacher led

a)

I drew the balloons. Then I crossed I out.

ASK

- Question ❶ a): *If you used counters here, what would each counter represent?*
- Question ❶ b): *What does crossing out represent?*

First there are 6 balloons.

Then I floats away.

Now there are 5 left.

I counted the balloons left.

IN FOCUS Question ❶ a) prompts children to associate '1 floats away' with crossing out a balloon. Dexter suggests how to find the answer: by counting the balloons that are left. Children can also refer to the number below the last balloon left. In this lesson, we are not introducing children to the subtraction symbol at this stage. This lesson focuses on children crossing out and counting how many are left.

b)

I more balloon floats away.

There are 4 balloons left.

STRENGTHEN In question ❶ b), ensure children can identify that the first balloon crossed out was from question ❶ a), and that one more has now floated away. Ask children how many balloons have gone in total.

121

PUPIL TEXTBOOK 1A PAGE 121

Think together

Whole class teacher led (I do, We do, You do)

ASK

- Question ❶: *How many balloons are there in total?*
- Question ❶: *How do you know how many have popped?*

IN FOCUS Question ❸ scaffolds children in writing their own number sentences to match the pictures provided. (Note that they should not need to count each individual total, as they should see that it stays the same each time.)

STRENGTHEN Use cubes to represent the total number of balloons. Take away cubes to represent the number of balloons that popped. Children could use different colours, or separate cubes laid out in different lines.

DEEPEN In question ❸, Ash prompts children to spot a pattern. Give children a set of blank scaffolds to fill in for each question and ask them to explain what happens to the number of balloons in total/that pop/that are left.

For even more depth, ask children to recreate the same pattern, but start by taking away 4 balloons and then take away 1 less balloon each time.

ASSESSMENT CHECKPOINT In question ❸, assess whether children write the number of balloons underneath or count in their head. Check if they stop counting at the crossed-out balloons when working out how many are left.

ANSWERS

Question ❶: Now there are 6 left.

Question ❷: There are 4 balloons left.

Question ❸ a): First there are 5 balloons.
Then 1 balloon pops.
Now there are 4 balloons.

Question ❸ b): First there are 5 balloons.
Then 2 balloons pop.
Now there are 3 balloons.

Question ❸ c): First there are 5 balloons.
Then 3 balloons pop.
Now there are 2 balloons.

Question ❸ d): First there are 5 balloons.
Then 4 balloons pop.
Now there is 1 balloon.

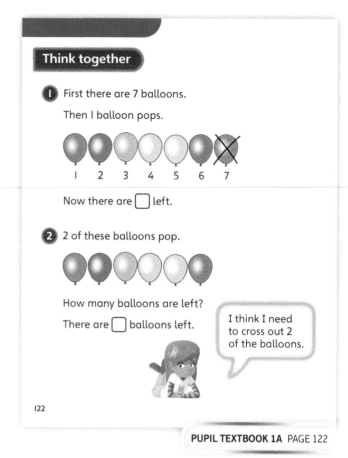

PUPIL TEXTBOOK 1A PAGE 122

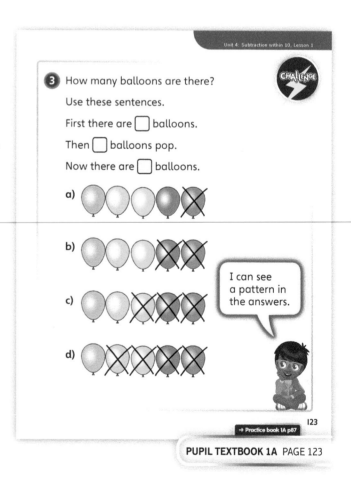

PUPIL TEXTBOOK 1A PAGE 123

Practice

WAYS OF WORKING Independent thinking

IN FOCUS Question ❶, provides children with pictorial representations of a subtraction, with snowmen, apples and candles crossed out. The scaffolding is reduced for question ❷ where children have to decide how many apples need crossing out.

STRENGTHEN Encourage children to use cubes or other concrete resources to represent each question.

DEEPEN In question ❹, children can work systematically from 10 birds in total and keep taking one away. Suggest that they try to find the same pattern as they did when doing the same thing with balloons.

THINK DIFFERENTLY The scaffolding is reduced in question ❺, where children are presented with a number problem to solve, within a real-life context. They need to decide what the key pieces of information are from the text in order to work out the answer. Although children could be directed to draw cars, they could use counters or shapes to represent the cars instead.

ASSESSMENT CHECKPOINT Ascertain whether children, when crossing out objects, cross out systematically or pick any particular pictures. Encourage crossing out from the right, as this is consistent with counting back on a number line.

ANSWERS Answers for the **Practice** part of the lesson can be found in the *Power Maths* online subscription.

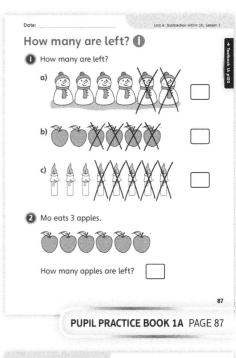

PUPIL PRACTICE BOOK 1A PAGE 87

PUPIL PRACTICE BOOK 1A PAGE 88

Reflect

WAYS OF WORKING Pair work, Whole class

IN FOCUS The **Reflect** question can prompt discussion about how to work out how many are left when some objects are taken away. Draw out that when some are taken away, the number left is less than the start number.

ASSESSMENT CHECKPOINT Assess how children explain the way in which they work out how many are left. Do they use phrases such as 'count back' or 'take away'? Do they explain that they took away a number of objects and then counted the remaining ones from 0, or do they explain counting back from the total?

ANSWERS Answers for the **Reflect** part of the lesson can be found in the *Power Maths* online subscription.

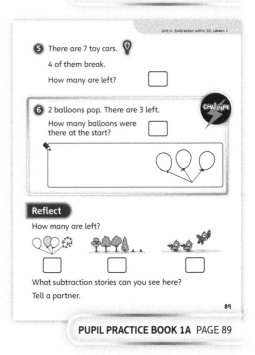

PUPIL PRACTICE BOOK 1A PAGE 89

After the lesson ⏸

- Were children comfortable with all of the different contexts for taking away and did they understand what the different numbers represented?
- Could children recognise and explain what approach they were taking to finding out how many were left?

How many are left? ❷

Learning focus

In this lesson, children will work out simple 'how many are left?' subtractions within 10 by using part-whole models and ten frames. They will use the minus symbol also.

Before you teach

- Are there any remaining misconceptions from Lesson 1 about counting how many are left that need to be addressed?
- Are all children secure with how to use ten frames to represent subtraction problems?

NATIONAL CURRICULUM LINKS

Year 1 Number – addition and subtraction

Represent and use number bonds and related subtraction facts within 20.

ASSESSING MASTERY

Children can solve subtractions within 10 in different contexts. Children can represent subtractions using different models, such as a ten frame or part-whole model, and differentiate between the number left and the number taken away. In this lesson, children will also be introduced to the minus symbol for subtraction.

COMMON MISCONCEPTIONS

Children may incorrectly interpret ways of representing a total number on a ten frame, not recognising the number taken away and the number left. Guide children towards strategies such as using different-coloured cubes, keeping separate piles of cubes or crossing out. Ask:
- *Which part represents the total number to start with?*
- *Which part represents the number taken away?*
- *Which part represents the number that is left?*

STRENGTHENING UNDERSTANDING

Use real-life stories to exemplify taking away. Encourage children to make up their own stories but always link them to the abstract calculation. Stories without writing the number sentences will not help children.

GOING DEEPER

Children start to spot which patterns they can find when answering questions. For example, ask: *If one more bird flies away, what happens to the answer? Is it one more or one less?*

KEY LANGUAGE

In lesson: subtract, take away, minus, left

Other language to be used by the teacher: represent

STRUCTURES AND REPRESENTATIONS

Ten frame, part-whole model

RESOURCES

Mandatory: counters, multilink cubes, ten frames

 In the eTextbook of this lesson, you will find interactive links to a selection of teaching tools.

Quick recap

Draw or make a first, then, now story for subtraction, for example, draw 5 apples and cross 2 out. Ask children to tell the first, then, now story.

Discover

How many are left? ❷

WAYS OF WORKING Pair work

Discover

ASK

- Question ❶ a): *Looking at the picture, can you point to each child in the classroom? Can you point to the two children leaving the classroom?*
- Question ❶ b): *Can you link the number sentence to the picture? What does the 8 represent? What does the 2 represent? What does the 6 represent?*

IN FOCUS In question ❶ a), children have the opportunity to explain how the different stages of the problem work: how many children there are in total to begin with, how many leave, and how many remain. In question ❶ b), children should be able to explain the meaning of the numbers and signs in the subtraction calculation in terms of the real context and the stages of the calculation.

PRACTICAL TIPS Act out the scene with eight children in the class whilst the others watch. Then swap so that all the children get a chance to act out the scene.

ANSWERS

Question ❶ a): There are 6 children left.

Question ❶ b): 8 – 2 = 6

8 is the number of children to start with.
2 is the number of children who go out.
6 is the number of children who are left.

❶ **a)** First there are 8 children.

Then 2 children go out of the room.

How many children are left?

b) Talk about the number sentence.

8 – 2 = 6

> Subtract can mean take away. The sign means minus (–).

124

PUPIL TEXTBOOK 1A PAGE 124

Share

WAYS OF WORKING Whole class teacher led

ASK

- Question ❶ a): *Why have we taken 2 counters away?*
- Question ❶ b): *What do the different parts of the calculation represent? What does the 8 mean? What does the – sign mean? What does the 2 mean? So, what does – 2 mean? What does the 6 represent?*

IN FOCUS In question ❶ a), children represent the problem using counters. It is important that they understand what the counters represent in relation to the number sentence. In question ❶ b), explain to children that we read 8 – 2 as 8 minus 2 and it is an example of a subtraction.

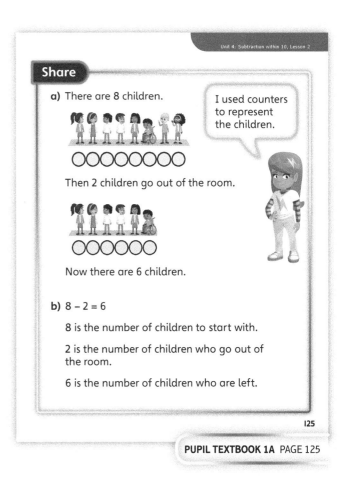

Share

a) There are 8 children.

○○○○○○○○

> I used counters to represent the children.

Then 2 children go out of the room.

○○○○○○

Now there are 6 children.

b) 8 – 2 = 6

8 is the number of children to start with.

2 is the number of children who go out of the room.

6 is the number of children who are left.

125

PUPIL TEXTBOOK 1A PAGE 125

Think together

WAYS OF WORKING Whole class teacher led (I do, We do, You do)

ASK

- Question ❶: *What does the 5 represent? What does the 2 represent? Will the answer be bigger or smaller than 5?*
- Question ❷: *Does it matter which way around the 7 and the 3 are in the number sentence? Which part of the subtraction is the total? Which sign, in the calculation, represents some being taken away? Which part of the subtraction shows what is left?*
- Question ❸: *What does the minus symbol mean? How can you use the ten frame to help you work out these number sentences?*

IN FOCUS In this part of the lesson, children can strengthen their understanding that the different parts of a subtraction calculation have meaning, in the given context. Children will also begin to write simple subtraction number sentences. In question ❸, children will put their understanding of the minus symbol into practice. They notice that the number of counters they remove from the ten frame is the number that follows after the minus symbol.

STRENGTHEN In question ❷, ask children to explain what each part of the subtraction represents as they are writing the numbers.

DEEPEN Can children subitise their answers for question ❸, when they take the counters away? For example, in part a) can they tell there are 7 counters left without counting?

ASSESSMENT CHECKPOINT Assess whether children put the – sign before or after the number being taken away. There may be some confusion because when we say, for example, '4 children leave', we say the '4' first, followed by the word denoting subtraction.

ANSWERS

Question ❶: 5 – 2 = 3

Question ❷: 7 – 3 = 4

Question ❸ a): 10 – 3 = 7

Question ❸ b): 10 – 4 = 6

Question ❸ c): 10 – 5 = 5

Question ❸ d): 10 – 6 = 4

PUPIL TEXTBOOK 1A PAGE 126

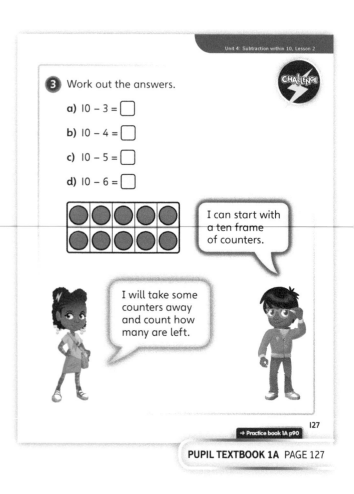

PUPIL TEXTBOOK 1A PAGE 127

Practice

WAYS OF WORKING Independent thinking

IN FOCUS Questions ❶, ❷ and ❸ provide pictorial representations of the subtractions, as well as the abstract calculations with missing numbers, to help children link the two representations together. In question ❸ b), prompt children to recognise that, in 8 – 4 = 4, the 4s have different meanings: the 4 that are subtracted are the birds that fly away. The 4 in the answer are the 4 that are left.

STRENGTHEN After question ❷, ask children to come up with their own sentences to match the pictures. For example: 'There were 8 apples. 3 were eaten. How many apples were left?'

DEEPEN Deepen understanding by encouraging children to come up with their own contexts to match the number sentences in question ❺.

THINK DIFFERENTLY In question ❹, children use full ten frames to calculate subtractions by crossing out the subtracted number of counters. This provides an opportunity to relate these subtractions to their knowledge of number bonds to 10.

ASSESSMENT CHECKPOINT Assess whether children are completing the subtraction sentences in questions ❸ and ❺ correctly. Do they understand that the start number is first, then the subtracted number, with the number left coming after the equals sign?

ANSWERS Answers for the **Practice** part of the lesson can be found in the *Power Maths* online subscription.

Reflect

WAYS OF WORKING Pair work

IN FOCUS This question asks children to describe a number fact in their own words (likely in a sentence). They could use cubes to help them represent it first, as Flo suggests.

ASSESSMENT CHECKPOINT Listen to children explaining the calculation. Do they use the phrases '5 in total' and 'take away 2'? Do they explain '– 2' by making the calculation concrete, saying 'cross two out' or 'take away two cubes'?

ANSWERS Answers for the **Reflect** part of the lesson can be found in the *Power Maths* online subscription.

After the lesson

- Did children understand the use of the – sign and what it meant in a subtraction?
- Were children able to understand subtractions in the pictures, concrete resources and different contexts?
- Did children use the language 'take away' correctly, both in context and alongside the abstract subtractions?

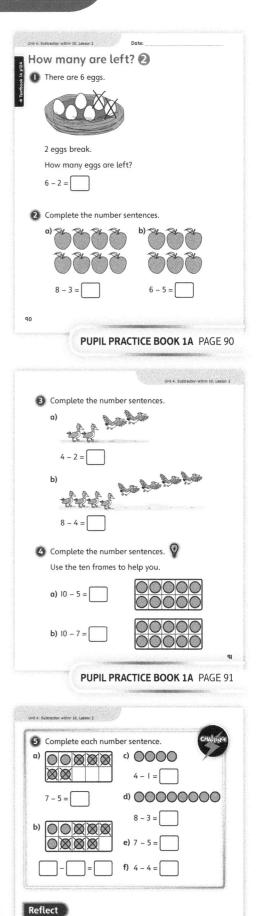

PUPIL PRACTICE BOOK 1A PAGE 90

PUPIL PRACTICE BOOK 1A PAGE 91

PUPIL PRACTICE BOOK 1A PAGE 92

165

Break apart **❶**

Learning focus

In this lesson, children will find two parts of a whole by breaking up a total. They will find one part by thinking about the whole and the other part.

Before you teach

- Are children confident in identifying the parts and the whole?
- If any children are unsure how to work out a missing part, how can this lesson be scaffolded to support them?

NATIONAL CURRICULUM LINKS

Year 1 Number – addition and subtraction

Represent and use number bonds and related subtraction facts within 20.

ASSESSING MASTERY

Children can break a whole into two parts and can reason about one part and the whole in order to work out a missing part. They are able to represent real-life situations using a part-whole model.

COMMON MISCONCEPTIONS

Children may confuse one of the parts for the whole or the whole for one of the parts. They may try to work out a missing part by adding the two numbers they can see. Ask:

- *The number of things you have to begin with, before you break it into parts, is called the whole. Where would you put this on the part-whole model? What does each number in the part-whole model represent?*

STRENGTHENING UNDERSTANDING

Spend time checking that children can identify the parts and the whole in a part-whole model. Use double-sided counters on ten frames to represent the parts so that children can easily see how the whole is made up.

GOING DEEPER

Deepen understanding by providing concrete examples of problems and using physical resources to represent objects in a problem. Put cubes or counters in a straight line, to emulate a number line. The number counted back can be collected and form a group, and the number left can become a different group. These can then be put into a blank part-whole model to represent the problem.

KEY LANGUAGE

In lesson: parts, whole, break

Other language to be used by the teacher: remain, in total, begin with, count back

STRUCTURES AND REPRESENTATIONS

Part-whole model, ten frame

RESOURCES

Mandatory: blank part-whole models, cubes, double-sided counters

Optional: toy cars

 In the eTextbook of this lesson, you will find interactive links to a selection of teaching tools.

Quick recap 🔄

Give children between 5 and 10 double-sided counters. Ask them to find all the bonds to this number and then represent the different bonds using the counters. How many do they know off by heart?

Discover

Break apart ❶

Discover

WAYS OF WORKING Pair work

ASK

- Question ❶ a): *How many cars are there in total? How will you show that 4 is a part and 9 is the whole?*

IN FOCUS Question ❶ a) gives you the opportunity to guide children towards thinking of 4 as one part of 9 and 9 as the whole. In question ❶ b), you can check children's understanding of the part-whole model before moving on.

PRACTICAL TIPS Use some toy cars to represent the situation or, if you do not have these, use cubes to represent the cars. Pretend the spaces on a ten frame are the car parking spaces.

ANSWERS

Question ❶ a): 9 is the whole. 4 is the part.

Question ❶ b): 5 is the other part.

❶ **a)** There are 9 cars.

4 of the cars are for sale.

What is the whole? What is a part?

b) What is the other part?

Draw the part-whole model.

128

PUPIL TEXTBOOK 1A PAGE 128

Share

WAYS OF WORKING Whole class teacher led

ASK

- Question ❶ a): *How many cars are there in total? Is this the whole or a part? How many cars are for sale? Is this the whole or a part?*
- Question ❶ b): *How can we find the missing part?*

IN FOCUS In question ❶ b), children relate the problem to a part-whole model. They have seen this model before. This time, however, the unknown is one of the parts and not the whole. Use children's knowledge of number bonds and addition at this stage rather than subtraction.

PUPIL TEXTBOOK 1A PAGE 129

Think together

WAYS OF WORKING Whole class teacher led (I do, We do, You do)

ASK

- Question ❶: *How many cubes are there in total? How many are Tim's? How many are Kat's? How can we represent this using a part-whole model?*
- Question ❷: *What could we use to represent the apples? How many apples have a leaf? How many apples have no leaf?*

IN FOCUS This section focusses on partitioning a whole into two parts. In questions ❶ and ❷, the part-whole model is provided to support children in linking a part-whole model with a contextual problem. In question ❸, children move to dealing with abstract calculations. As they work through the examples, children should be able to see that, if they know one part, then they can work out the other part. Encourage children to use concrete objects to help them, such as cubes and counters, and they may split them into two groups – this is fine. In Question ❸, children may use their knowledge of number bonds to help them.

STRENGTHEN In question ❷, the use of real apples (either with and without a leaf, or red and green apples) can help children see the parts. In question ❸, use double-sided counters to represent each problem.

DEEPEN Children could solve problems with two missing parts. How many ways can they find to solve the problems? Encourage children to use their knowledge of number bonds where they can, particularly in question ❸.

ASSESSMENT CHECKPOINT Children can find missing parts in part-whole models and abstract calculations. They can represent problems using double-sided counters or different-coloured cubes

ANSWERS

Question ❶: $8 - 5 = 3$
3 of the cubes are Kat's.

Question ❷: $7 - 2 = 5$
5 apples have no leaves.

Question ❸ a): 2

Question ❸ b): 3

Question ❸ c): $4 + \mathbf{2} = 6$
$3 + \mathbf{3} = 6$
$\mathbf{5} + 1 = 6$

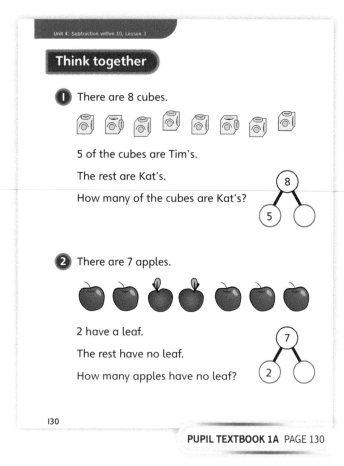

PUPIL TEXTBOOK 1A PAGE 130

PUPIL TEXTBOOK 1A PAGE 131

Practice

WAYS OF WORKING Independent thinking

IN FOCUS Questions progress in terms of a reduction in scaffolding. Questions ❶ and ❷ provide pictorial representations and partially filled part-whole models, question ❸ shows only partially filled part-whole models, and question ❹ moves onto abstract representations of subtractions. Encourage children to use knowledge of bonds or concrete materials as necessary. The focus of this practice is very much on finding the missing part. Use this language throughout and avoid saying 'How many are left?'.

STRENGTHEN For all questions, use double-sided counters or two different colours of cubes as needed to support children.

DEEPEN Deepen understanding by asking children to come up with their own contexts to match the part-whole models in question ❸. Children will need to come up with contexts in which two groups make an overall total, such as bananas and apples making up a bowl of fruit.

ASSESSMENT CHECKPOINT Assess how children approach each new question. Do they have consistent strategies for working out the wholes and/or parts? Do they circle wholes/parts or draw the calculations themselves? Do they rely on previous part-whole knowledge and/or use resources to check?

ANSWERS Answers for the **Practice** part of the lesson can be found in the *Power Maths* online subscription.

Reflect

WAYS OF WORKING Independent thinking

IN FOCUS This question provides a missing number problem, framed in a slightly different way; using a subtraction instead of an addition (children tackled missing number problems with addition in question ❹).

ASSESSMENT CHECKPOINT Do children recognise that the focus is still on finding the missing part, even though it involves a subtraction, and they can solve it in the same way they have done with the empty boxes and part-whole models throughout the lesson? They will tackle problems like these in the next lesson.

ANSWERS Answers for the **Reflect** part of the lesson can be found in the *Power Maths* online subscription.

After the lesson ⏸

- Were children able to use their knowledge of parts and wholes to solve missing number problems?
- How confident were children with their number bonds within 10?
- Were children able to represent problems using concrete manipulatives?

PUPIL PRACTICE BOOK 1A PAGE 93

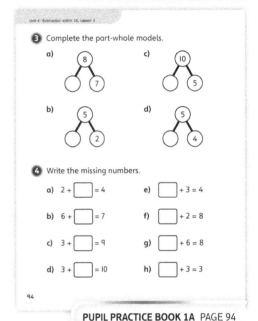

PUPIL PRACTICE BOOK 1A PAGE 94

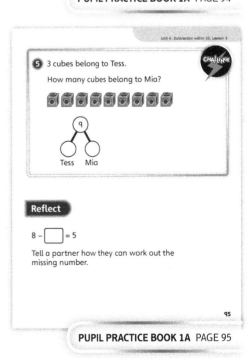

PUPIL PRACTICE BOOK 1A PAGE 95

Break apart ❷

Learning focus

In this lesson, children will continue to find a missing part from the whole and the other part. Children will use a part-whole model and sentence scaffolds to help them to complete subtractions.

Before you teach

- Are all children secure with relating a part-whole model to a subtraction?
- Are children able to spot the mistake if a part and a whole are put into the wrong places in a part-whole model?

NATIONAL CURRICULUM LINKS

Year 1 Number – addition and subtraction

Represent and use number bonds and related subtraction facts within 20.

ASSESSING MASTERY

Children can correctly put a whole and a part from a subtraction into a part-whole model, recognising that the missing part forms a part of the whole that can be worked out using the remaining numbers. Children can start to link this process with subtraction calculations (whole – part = part).

COMMON MISCONCEPTIONS

When translating a subtraction, in which a part is missing, into a part-whole model, children may put both remaining numbers into the parts positions (as they have seen when learning about number bonds). Ask:
- *Does the subtraction tell you the whole? Where would the whole go on the part-whole model?*

STRENGTHENING UNDERSTANDING

Strengthen understanding by asking children to work in pairs, taking turns to show a number of cubes or counters to their partner and then hiding some (in their hand/under the table). The partner should then calculate the quantity missing. Use a container with defined sections for resources (such as a blank ten-frame or an egg box), so children can count the empty sections.

GOING DEEPER

Deepen understanding of subtractions by asking children to check them using addition. First, ask children to put the subtraction into a part-whole model. Ask: *Can you recognise which two numbers need to be added together to make the whole?* Provide blank calculations to help.

KEY LANGUAGE

In lesson: subtraction, addition, missing part, minus

Other language to be used by the teacher: remain, in total, begin with, count back

STRUCTURES AND REPRESENTATIONS

Part-whole model

RESOURCES

Mandatory: blank part-whole models, multilink cubes, counters

Optional: blank ten frames, empty egg boxes, real or play fruit

 In the eTextbook of this lesson, you will find interactive links to a selection of teaching tools.

Quick recap 🔎

Give children a rod of between 5 and 10 cubes. Ask them to say how many cubes they have. Ask them to break their rod into two parts. Ask: *What is the whole? What are the parts?* How many different ways can they break the whole?

Discover

Break apart ❷

WAYS OF WORKING Pair work

ASK

- Question ❶ a): *How many counters can you see in total, in the first pair of hands?*
- Question ❶ a): *When some counters are being hidden, has the total number changed?*
- Question ❶ a): *Without looking in the hand, how can we work out how many counters are hidden?*

IN FOCUS This part of the lesson prompts children to work out a subtraction by thinking about how the parts relate to the whole. Ask children to look closely at the picture and to tell you what they see. Then read out the statements in question ❶ a), and ask children to point to the part of the picture that relates to each statement.

PRACTICAL TIPS Recreate the scenario in the classroom, with children working in pairs with each pair using five counters or coins.

ANSWERS

Question ❶ a): The other part is 3.

Question ❶ b):

Discover

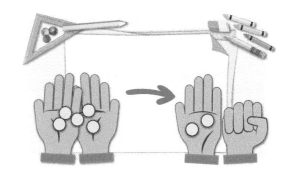

❶ **a)** 5 is the whole.

   2 is a part.

   How many counters are in the other part?

b) Draw a part-whole model.

132

PUPIL TEXTBOOK 1A PAGE 132

Share

WAYS OF WORKING Whole class teacher led

ASK

- In response to Astrid's comment: *Why don't you have to guess?*
- Question ❶ a): *How can you use counters to complete the missing part?*
- Question ❶ b): *What numbers are the parts? Which number is the whole?*

IN FOCUS Question ❶ b), tests children's understanding of the part-whole model and being able to identify a missing part from knowing the other part and the whole. Make sure that children understand what each number in the part-whole model represents in relation to the activity (e.g. the 5 represents the whole, 3 is a part and 2 is the other part). Explain how this relates to the work on bonds to 5 they did earlier in the book.

DEEPEN To deepen understanding, ask children what would have changed if 2 counters had been hidden instead of 3. Encourage them to create their own part-whole models to show the missing part. Ask: *Do you get a different answer?*

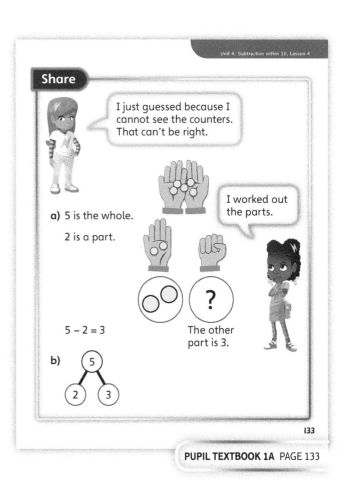

PUPIL TEXTBOOK 1A PAGE 133

Think together

Unit 4: Subtraction within 10, Lesson 4

Think together

WAYS OF WORKING Whole class teacher led (I do, We do, You do)

ASK
- Question **1**: *What do 5 and 4 represent? Which is a part and which is a whole?*
- Questions **1** and **2**: *How can the pictures help you work out what is missing?*
- Question **3**: *Can you show how many cakes are hidden using a different resource?*

IN FOCUS Questions **1** and **2** allow children to follow the same method as in **Share** to work out a missing part. Children should be increasing in confidence to find a missing part when they know the whole and the other part. Question **3** has reduced scaffolding, to encourage children to decide how they are going to approach the question. As well as using cubes or drawing circles to represent the cakes, they could also draw a part-whole model and write a number sentence.

STRENGTHEN Questions **1** and **2** are perfect examples to practise using counters in pairs. Ask children to act out the situation. Get them to do more types of problems like this. It is okay at this stage if children have to count the remaining counters. Strengthen understanding by making the context of question **3** concrete: ask children to act out the problem using different resources.

DEEPEN For question **3**, ask children to create their own addition to check how many cakes are hidden. Encourage them to label each part of their addition with 'cakes hidden under the cloth', 'cakes I can see' and 'total number of cakes'.

ASSESSMENT CHECKPOINT For questions **1** and **2**, ensure children understand that it does not matter in which part section of the part-whole model the known part is written. In question **3**, assess whether children draw and then cross out or circle cakes when solving the subtraction problem, as they have seen done in previous lessons.

ANSWERS

Question **1**: $5 - 4 = 1$

Question **2**: $7 - 4 = 3$

Question **3**: $6 - 2 = 4$
    4 cakes are hidden.

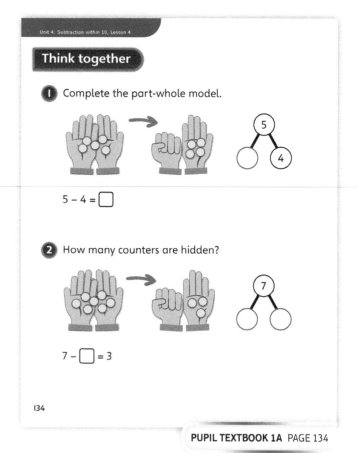

Think together

1 Complete the part-whole model.

$5 - 4 = \boxed{}$

2 How many counters are hidden?

$7 - \boxed{} = 3$

134

PUPIL TEXTBOOK 1A PAGE 134

3 There are 6 cakes in total.

How many cakes are hidden?

CHALLENGE

I can draw I counter for each cake.

I can use cubes.

To check a subtraction you can do an addition.

135

→ Practice book 1A p96

PUPIL TEXTBOOK 1A PAGE 135

Practice

WAYS OF WORKING Independent thinking

IN FOCUS In question ①, children are finding missing parts in part-whole models. They should see how the subtraction sentence helps them work out the missing number. In questions ② and ③, children have problems in word format and they have to find the missing part. They should think about this as a part-whole model. What is the whole and what is one of the parts? Encourage children to use their knowledge of number bonds to help them find the missing numbers. In question ⑤, children are asked to complete abstract subtractions, using part-whole models to help them. Before children do any calculations for question ⑤ a), take this opportunity to ask: *Do you think the answer will be less than the whole and/or the other part?*

STRENGTHEN In question ①, ask children to use counters or cubes, similar to the **Share** section in the **Textbook**, to work out the missing parts. In question ②, allow children to represent the word problem with real or play fruit or coloured counters. Although the subtraction could be 8 – 5 = 3 or 8 – 3 = 5, 8 – 5 = 3 is better in the context.

DEEPEN After question ②, encourage children to create their own subtractions for a partner to answer.

ASSESSMENT CHECKPOINT In questions ③ and ④, assess whether children can correctly represent the question as a subtraction sentence. Do they understand that the number that has been broken apart is the first number in the subtraction?

ANSWERS Answers for the **Practice** part of the lesson can be found in the *Power Maths* online subscription.

Reflect

WAYS OF WORKING Pair work

IN FOCUS The **Reflect** question targets the misconception of mixing up the whole and a part in subtractions. Children should emphasise that the whole must always be the first number in the subtraction sentence.

ASSESSMENT CHECKPOINT Assess whether children are able to identify the mistakes and articulate what has happened. Ask them to write the correct subtractions to match the part-whole models when they have explained which parts have been mixed up.

ANSWERS Answers for the **Reflect** part of the lesson can be found in the *Power Maths* online subscription.

After the lesson ⏸

- Were there sufficient opportunities within this lesson for children to experiment with resources and their own contexts to make subtractions?
- Have children mastered calculating subtractions and using corresponding part-whole models?
- Were children able to identify and articulate when mistakes with part-whole models have been made, and correct mistakes?

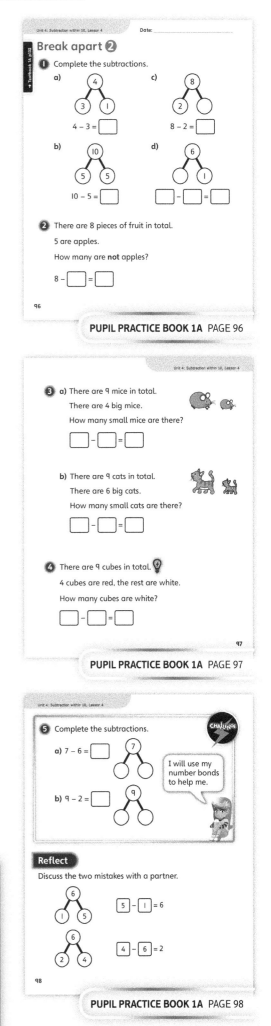

PUPIL PRACTICE BOOK 1A PAGE 96

PUPIL PRACTICE BOOK 1A PAGE 97

PUPIL PRACTICE BOOK 1A PAGE 98

Fact families

Learning focus

In this lesson, children will find four addition and four subtraction facts from the same context. They will see the = sign in different parts of number sentences and be able to explain the meaning of each part of the number sentences.

Before you teach

- If children have previously seen number sentences with the = sign only at the end, how will you first introduce its use in a different place?
- What resources could you use to help children to derive different number facts from a context?

NATIONAL CURRICULUM LINKS

Year 1 Number – addition and subtraction

Represent and use number bonds and related subtraction facts within 20.

ASSESSING MASTERY

Children can derive two addition and two subtraction facts from a context, understanding how to translate them into number sentences, and create number sentences to show a specific fact. Children can identify what each part in the number sentence means and switch numbers around to show the same fact.

COMMON MISCONCEPTIONS

Children may think that the = sign means 'write the answer here', rather than 'is equal to'. Ask:
- What does the = sign mean?

STRENGTHENING UNDERSTANDING

Strengthen understanding by using printed +, – and = signs and number cards, or magnetic numbers. Children can then physically move the numbers in a calculation around, or test out combinations of which they are unsure before writing them down. Children could also move numbers around pre-printed scaffolds, if you would prefer them to clarify what type of number sentences they are making.

GOING DEEPER

Once children are confident with deriving different number sentences from one context, deepen understanding by looking at the patterns they have made and coming up with generalised number sentences. For example:

part + part = whole   whole = part + part   part = whole – part

part + part = whole   whole – part = part   part = whole – part

whole = part + part   whole – part = part

KEY LANGUAGE

In lesson: subtract, addition, plus, minus, part, whole, **fact family**

Other language to be used by the teacher: remain, in total, begin with, count back

STRUCTURES AND REPRESENTATIONS

Part-whole model

RESOURCES

Mandatory: blank part-whole models, cubes and/or counters

Optional: printed +, – and = signs, number cards or magnetic numbers, pre-printed number sentence scaffolds, small hoops and a post or bean bags and a large hoop

 In the eTextbook of this lesson, you will find interactive links to a selection of teaching tools.

Quick recap

Write a completed part-whole model on the board. Ask children to write two addition facts that they can see from the part-whole model.

Discover

Fact families

Discover

WAYS OF WORKING Pair work

ASK

- Question ① a): *How does the picture relate to the facts you are given?*
- Question ① a): *What do the 6 and 2 represent? Can you see the 6? Can you see the 2?*
- Question ① b): *What number will represent the whole?*

IN FOCUS Question ① a) prompts children to calculate how many rings land on the post. This is a starting point for question ① b), which asks children to write four number sentences for this fact. This process builds on previous learning of building a fact family. It can also lead to a discussion of the = sign and and where it should appear in a number sentence.

PRACTICAL TIPS Recreate the scenario in the classroom using concrete resources, such as hoops on a post or bean bags in a hoop.

ANSWERS

Question ① a): 4 rings land on the post.

Question ① b): 6 – 2 = 4
6 – 4 = 2
2 + 4 = 6
4 + 2 = 6

① **a)** Leon throws 6 rings.

2 rings miss the post.

How many land on the post?

b) Write four different number sentences to show this fact.

136

PUPIL TEXTBOOK 1A PAGE 136

Share

WAYS OF WORKING Whole class teacher led

ASK

- Question ① b): *What do the numbers in the number sentences mean? Can you think of any other number sentences for this fact?*

IN FOCUS Question ① a) gives an opportunity to embed understanding of parts, a whole and the structure of a part-whole model. Suggest that, when looking at the part-whole model and number sentence, children assign each of the numbers the label 'part' or 'whole' and ascertain whether the part-whole model has been drawn correctly. Children should be using their knowledge of all of their previous work with part-whole models, as well as addition and subtraction facts.

DEEPEN When looking at question ① a), ask children which of the numbers in the number sentence could be switched around to give the same fact. Discuss why 2 – 6 = 4, for example, does not work.

Share

a) There are 6 rings in total.

2 rings miss.

6 – 2 = 4

4 rings land on the post.

b)

I found four number sentences.

6 – 2 = 4
6 – 4 = 2
2 + 4 = 6
4 + 2 = 6

I think there are some more.

These four number sentences make up a fact family.

137

PUPIL TEXTBOOK 1A PAGE 137

175

Think together

Whole class teacher led (I do, We do, You do)

ASK

- Question **1**: *Which number is the whole? Which are the parts? Is the whole always after the = sign?*
- Question **1**: *How does the part-whole model relate to the picture of rings on the post?*
- Question **2**: *What could a picture for this part-whole model look like? Can you represent your picture with counters? What addition facts can you see? What subtraction facts can you see?*

IN FOCUS Questions **1** and **2** give children practice at writing two addition and two subtraction facts for each part-whole model, resulting in what is known as a fact family. Ask children why they think it is called a fact family. Question **3** begins to explore that you can get eight facts from a ten frame showing parts and wholes like this. Children may also see that they can write down a fact family if they know just one of the facts.

STRENGTHEN Use counters or cubes throughout to support children's understanding. Children may need more practice writing down a fact family for a part-whole model. Ask them to make their part-whole model and then challenge a partner to write down 4 facts.

DEEPEN Talk about what types of number sentence can be worked out from one context. Ask: *Are there more addition or more subtraction facts?* Draw children to the conclusion that there are an equal number of addition and subtraction facts because each is the inverse of the other. Discuss how number sentences can also be varied depending on where the = sign is. Isolate a 'whole – part' calculation and show that the '= missing part' can go either side, but the calculation does not change.

ASSESSMENT CHECKPOINT Assess whether children are able to explain what they are trying to find, and are correctly putting numbers into the number sentences.

ANSWERS

Question **1**: 3 + 4 = 7
4 + 3 = 7
7 – 3 = 4
7 – 4 = 3

Question **2**: 1 + 5 = 6
5 + 1 = 6
6 – 1 = 5
6 – 5 = 1

Question **3** a): 2 + 8 = 10; 10 = 2 + 8
8 + 2 = 10; 10 = 8 + 2
10 – 2 = 8; 8 = 10 – 2
10 – 8 = 2 ; 2 = 10 – 8

Question **3** b): 7 + 3 = 10; 10 = 7 + 3
3 + 7 = 10; 10 = 3 + 7
10 – 3 = 7; 7 = 10 – 3
10 – 7 = 3; 3 = 10 – 7

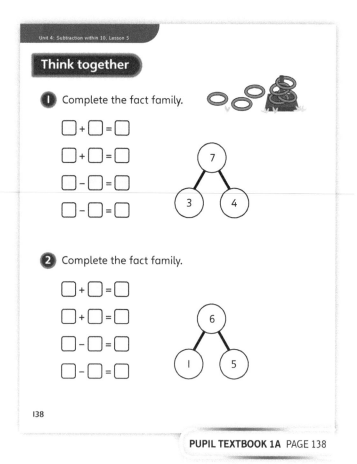

PUPIL TEXTBOOK 1A PAGE 138

PUPIL TEXTBOOK 1A PAGE 139

Practice

WAYS OF WORKING Independent thinking

IN FOCUS In question **1**, children are representing a real-life scenario in a part-whole model. This will help them with some of the later questions where the contexts are more abstract. They could use the frog scenario in other questions to help them make sense of them. In questions **2** and **3**, children further practice writing down the 4-fact family for each of the part-whole models. They should be able to link the abstract sentences to the pictures given. For example, they should be able to say that 3 + 2 = 5 in question **3**, and that the 3 and 2 represent the children on the ends of the see-saw and that 5 is the total. Question **4** asks children to write down the 8-fact family.

STRENGTHEN Consider completing the whole for the children in question **1**, to get them started. If they are still unsure, spend time supporting them using counters or cubes and physically put them together to show addition or break them apart to show subtraction.

DEEPEN If children are confident filling in different blank number sentence scaffolds from a context, encourage them to write the blank scaffolds themselves, representing missing numbers as squares and missing signs as circles. To further extend children, give them a sentence, such as 6 + 2 = 8. Can they write down the rest of the fact family? Ask children what the part-whole for this fact family might look like.

ASSESSMENT CHECKPOINT Assess whether children can confidently write down 2 addition and 2 subtraction sentences (i.e. the 4-fact family) for a given part-whole model or word style problem. Some children may be able to write down all 8 facts (where they see the = sign can be in a different position).

ANSWERS Answers for the **Practice** part of the lesson can be found in the *Power Maths* online subscription.

Reflect

WAYS OF WORKING Pair work

IN FOCUS The **Reflect** question prompts children to explain and reword what they have learnt this lesson, with no visual prompts or cues and no scaffolds to help them.

ASSESSMENT CHECKPOINT Assess whether children are able to generate three other subtractions from this fact or just one other, with the = sign at the end, and whether they understand how addition facts are linked to subtraction. Look for any children who still rely on drawing a part-whole model to help them think through what each number is.

ANSWERS Answers for the **Reflect** part of the lesson can be found in the *Power Maths* online subscription.

After the lesson

- Were all children able to write down two addition and two subtraction facts for the part-whole models?
- Were children able to relate addition and subtraction facts, and understand which numbers could be switched around to retain the same facts?
- Did children have to calculate at any point in the lesson to check their work, or did they trust their ability to arrange and rearrange number sentences correctly?

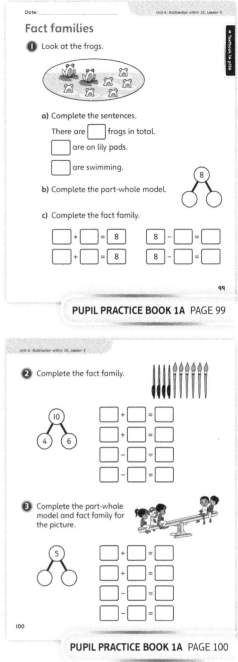

PUPIL PRACTICE BOOK 1A PAGE 99

PUPIL PRACTICE BOOK 1A PAGE 100

PUPIL PRACTICE BOOK 1A PAGE 101

Subtraction on a number line

Learning focus

In this lesson, children will calculate subtraction number sentences using a number line to count back from the bigger number.

Before you teach

- Are all children secure with using a number line when counting on?
- Are all children secure with the term 'less than'?

NATIONAL CURRICULUM LINKS

Year 1 Number – addition and subtraction

Solve one-step problems that involve addition and subtraction, using concrete objects and pictorial representations, and missing number problems such as $7 = \square - 9$.

ASSESSING MASTERY

Children can work out a subtraction number sentence by counting how many steps back on a number line they need to jump to arrive at the answer, recognising that the number of steps is one of the numbers in the subtraction. Children can relate the – to counting back.

COMMON MISCONCEPTIONS

When working out a subtraction using the method of counting back, children may count the number they are starting on. For example 9 – 3: children may count 9, 8, 7 as opposed to 8, 7 and 6. Ask:

- *Where they do you start?*
- *How jumps back are you making?*
- *How can you represent these jumps on the number line?*

To help avoid this misconception, encourage children to draw 'jumps' as arrows on a number line. Each jump represents a count back of 1.

STRENGTHENING UNDERSTANDING

Strengthen understanding of counting back by using a hopscotch grid outside or a giant number line in the classroom. Children can stand by their starting number and then physically jump back the correct number of times to see where they land.

GOING DEEPER

Once children are confident about counting back to find the answer to a subtraction number sentence, deepen understanding by asking them to use their knowledge of addition to check the calculation. Ask children to count on the same number of steps to check whether the same starting point is reached. If they can, ask them to give the addition number sentence that matches what they are doing.

KEY LANGUAGE

In lesson: count back, less than, subtract, take away

Other language to be used by the teacher: starting number

STRUCTURES AND REPRESENTATIONS

Number line

RESOURCES

Mandatory: number lines, blank number lines, multilink cubes, counters, 0 to 10 number tracks, dice

Optional: hopscotch grid, giant number line

 In the eTextbook of this lesson, you will find interactive links to a selection of teaching tools.

Quick recap

Give children a 0 to 10 number track, a counter and a dice. Start with the counter on 10. Ask children to roll the dice and move their counter back that many places. What number do they land on? Can they reach 0? How many rolls does it take?

Discover

Subtraction on a number line

Discover

WAYS OF WORKING Pair work

ASK

- Question ① a): *Where has Maya started on the number line?*
- Question ① a): *Is she jumping on or back?*
- Question ① a): *When Maya has jumped, what is a strategy to keep track of what her starting number was? Circle it.*

IN FOCUS These questions use a pictorial representation of a number line. Ask children to identify what Maya is jumping on in the picture. They also require children to understand that the term 'more' does not necessarily suggest addition: when saying 'Maya jumps two *more* times', ask children: *In which direction will she be jumping?* Point out that it is the number of steps Maya is jumping *back* that is getting bigger, not the number on the number line.

PRACTICAL TIPS Recreate the scenario outside on a hopscotch grid or using a giant number line in the classroom.

ANSWERS

Question ① a): Maya lands on 6.

Question ① b): Maya lands on 4.

① a) Maya has to jump 3 times ⌐3⌐.
What number will she land on?

b) Maya jumps 2 more times ⌐2⌐.
Where does she land?

140

PUPIL TEXTBOOK 1A PAGE 140

Share

WAYS OF WORKING Whole class teacher led

ASK

- Question ① a): *Look at how Dexter has counted back. Has he started at the right number? Has Maya started at the right number?*
- Question ① a): *Do we need to find 3 on the number line?*
- Question ① b): *When Maya jumps back 2 more times, what was her starting point?*

IN FOCUS These questions allow you to ensure children understand they should not start their count back with their starting number, and that 3 in this context is the number of jumps back Maya makes and is not the 3 already showing on the number line.

DEEPEN Ask children how many jumps back Maya has done in total, from 9 to 4. They could use their own number lines to help them count how many jumps back they have to make. Some children may spot that the 5 total jumps back are made from the parts 3 and 2, and they will be able to write the subtraction number sentence 9 − 5 = 4 to match all of the jumps back.

Share

I can count back:
9, 8, 7, 6, 5, 4, 3, 2, 1, 0.

a) 9 − 3 = 6

8...7...6

0 1 2 3 4 5 6 7 8 9 10

Maya lands on 6.

b)

0 1 2 3 4 5 6 7 8 9 10

Maya starts on 6.

She jumps 2 times...5...4.

6 − 2 = 4

Maya lands on 4.

141

PUPIL TEXTBOOK 1A PAGE 141

Think together

WAYS OF WORKING Whole class teacher led (I do, We do, You do)

ASK

- Question ❶: *Where does the number sentence show '2 less'?*
- Question ❶: *What does this look like on the number line?*
- Question ❸: *Where is 0 on each of these number lines?*

IN FOCUS In questions ❷ and ❸, each subtraction sentence is presented with a corresponding number line. Children should understand that they should identify the starting point and the number of jumps back they should make to find the answer. This is different to the crossing-out method they did earlier in this unit. Children should increase in confidence by knowing where they start and how many they have to count back. They realise that the number they land on is the answer to the subtraction. Some children may be able to use their number bonds too. In question ❸, all of the number sentences have the same answer: '0'. This provides an opportunity for you to guide children towards the understanding that any number taken away from itself will be 0, referring to what Ash says about seeing a pattern.

STRENGTHEN Strengthen children's understanding by asking them to consider patterns in what they see. Ask: *When counting back on a number line, is the number reached always less than the starting number? Does that mean that the starting number is always more than the number reached?* Ask children to explain, using the numbers and examples they have just seen.

DEEPEN Ask children to show the pattern in question ❸ using a different resource. For example, 3 cubes take away 3 cubes gives 0 cubes. Ask children to come up with more examples of number sentences with the answer 0. Encourage them to express what generalised pattern they are seeing in a sentence.

ASSESSMENT CHECKPOINT Assess whether children recognise that the number of jumps back on a number line is one of the numbers in the subtraction.

ANSWERS

Question ❶: 6 – 2 = 4; Maya lands on 4

Question ❷: 7 – 3 = 4
6 – 4 = 2
9 – 5 = 4

Question ❸ a): 3 – 3 = 0

Question ❸ b): 2 – 2 = 0

Question ❸ c): 1 – 1 = 0

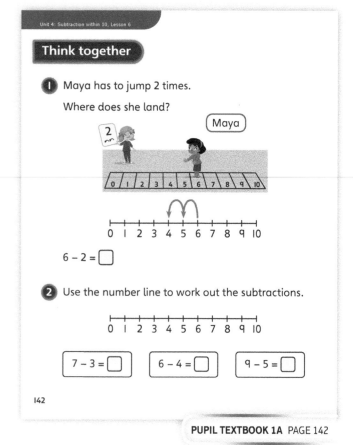

PUPIL TEXTBOOK 1A PAGE 142

PUPIL TEXTBOOK 1A PAGE 143

Practice

IN FOCUS In question ③, children need to make the connection between counting back and the subtraction number sentence. In questions ① and ②, children need to count back onto spaces presented differently from a regular number line. In question ⑤, children have to use the number line to work out how many were subtracted if they know where they ended.

STRENGTHEN In question ⑤, children have to generate different subtraction number sentences with the same answer. Strengthen children's grasp of the question by suggesting they work systematically to find all the different subtractions, starting from the answer and counting on to find a starting point greater than it.

DEEPEN Ask children to complete the question ⑤ activity for different target numbers under 10. Encourage them to spot a pattern between the target number and the number of number sentences that can be generated for it (with the maximum starting number being 10): there will always be one more number sentence than the target number, because the operation '− 0' should be included.

ASSESSMENT CHECKPOINT Assess whether children, when counting back, make the mistake of starting the count from their starting number. When they count back out loud, ensure children say out loud the numbers on which they land on a number line, not the number of jumps: if they are counting back 2 from 7, for example, they should say, 'Six, five'. They also need to recognise when to stop the count.

ANSWERS Answers for the **Practice** part of the lesson can be found in the *Power Maths* online subscription.

Reflect

IN FOCUS The **Reflect** question allows children to explain and reword different subtraction strategies, such as using part-whole models, counting what is left, breaking apart a whole and finding the difference on a number line.

ASSESSMENT CHECKPOINT Assess whether children can explain each method. Listen to assess whether there is a method children prefer.

ANSWERS Answers for the **Reflect** part of the lesson can be found in the *Power Maths* online subscription.

After the lesson

- Did children understand what the different parts of subtraction number sentences represent on a number line? For example, did they understand that the part being taken away represents the number of jumps?
- Were children confident using 0 in number sentences: counting back to 0 and counting back 0 jumps?

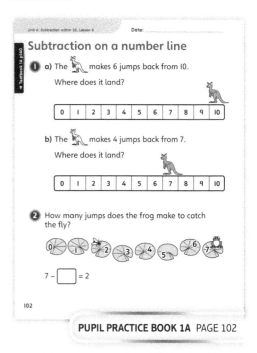

PUPIL PRACTICE BOOK 1A PAGE 102

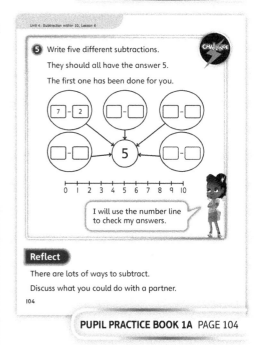

PUPIL PRACTICE BOOK 1A PAGE 103

PUPIL PRACTICE BOOK 1A PAGE 104

Add or subtract 1 or 2

Learning focus

In this lesson, children will add or subtract 1 or 2 using a number line to help them. They will draw on their knowledge of addition and subtraction from previous lessons in the unit.

Before you teach

- Do children understand the difference between addition and subtraction?
- Are children confident counting on and back on a number line?

NATIONAL CURRICULUM LINKS

Year 1 Number – addition and subtraction

Add and subtract 1-digit and 2-digit numbers to 20, including 0.

ASSESSING MASTERY

Children can add and subtract 1 or 2. Although a number line is used for support, children should be encouraged to work these out mentally. They should start trying to picture the number line in their heads.

COMMON MISCONCEPTIONS

When counting on or back to add or subtract 1 or 2, children may include the number they are starting at. For example, when solving 6 – 2 they may count '6, 5' and think the answer is 5 because they have said two numbers. Ask:
- *What is the starting number? How many are being taken away from the starting number?*

STRENGTHENING UNDERSTANDING

Use counters and a ten frame to remind children of adding more and taking away. Do this alongside a number line to reinforce the way children are counting on and back. Once they have shown a subtraction on a ten frame, such as 6 – 2, ask them what this would look like on a number line.

GOING DEEPER

Encourage children to solve problems mentally and describe how they have done it. What picture can they see in their head?

KEY LANGUAGE

In lesson: add, subtract, take away

STRUCTURES AND REPRESENTATIONS

Number line, ten frame

RESOURCES

Mandatory: blank number lines, digit cards

Optional: counters, cubes

 In the eTextbook of this lesson, you will find interactive links to a selection of teaching tools.

Quick recap

Make some digit cards from 1 to 9. Ask children to choose a digit card. What is 1 more than the number? What is 1 less than the number? Repeat for other numbers.

Discover

Add or subtract 1 or 2

Discover

Amy

WAYS OF WORKING Pair work

ASK

- Question ① a): *Can you point to 6 on the number line? If you add 1, will your answer be bigger or smaller than 6?*
- Question ① b): *If you add 2, will your answer be bigger or smaller than adding 1?*

IN FOCUS In this section, children see that when adding to a number, they count on to the right of the start number. They will learn that adding 1 or 2 to a number makes it bigger.

PRACTICAL TIPS A ten frame and counters could also be used alongside the number line so that children can check their answer in two ways.

ANSWERS

Question ① a): 6 + 1 = 7

Question ① b): 6 + 2 = 8

① **a)** Amy has put 6 apples in a line.

If she adds 1 apple to the line, how many apples will there be?

Use a number line to work out 6 + 1.

b) If she adds 2 apples to the line, how many will there be?

144

PUPIL TEXTBOOK 1A PAGE 144

Share

WAYS OF WORKING Whole class teacher led

ASK

- Question ① a) and ① b): *Why have we started at 6 on the number line? How many jumps have we made on the number line? Why?*

IN FOCUS Children see the difference between adding 1 and adding 2. They should be able to relate the start number and the number of jumps on the number line to the question. Encourage children to use their fingers, as in the **Textbook**, showing where they start and where they count on to. They should realise that + 1 is the same as one more and + 2 is one more than that.

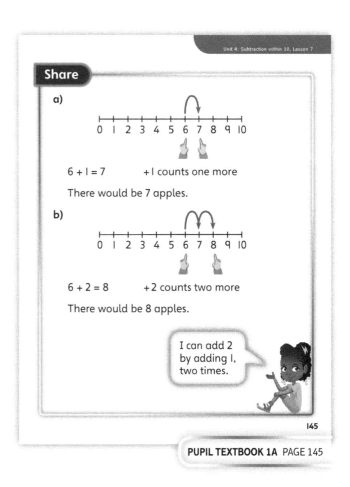

Share

a)

6 + 1 = 7   + 1 counts one more

There would be 7 apples.

b)

6 + 2 = 8   + 2 counts two more

There would be 8 apples.

I can add 2 by adding 1, two times.

145

PUPIL TEXTBOOK 1A PAGE 145

183

Think together

WAYS OF WORKING Whole class teacher led (I do, We do, You do)

ASK

- Questions **1** a): *Why are you pointing to 5 on the number line? Should you count on or back?*
- Questions **1** b): *Will the answer be bigger or smaller than your answer to part a?*

IN FOCUS In question **1**, children are asked to subtract 1 and 2 using a number line. Children can see that, when subtracting, they need to count back, not on. Questions **2** and **3** ask children to use a number line for adding and subtracting 1 and 2. By attempting the **Think together** questions, children are strengthening their ability to add and subtract 1 and 2. They will deepen their understanding of how this relates to '1 more' and '1 less'.

STRENGTHEN Use a ten frame and counters to remind children what subtraction means. Then relate this to what they need to do on the number line.

DEEPEN Reverse the problem for children who need a challenge. For example, 'When I add 2, the answer is 8. What is my starting number?'

ASSESSMENT CHECKPOINT Children can confidently add or subtract 1 or 2 and show this on a number line with the appropriate number of jumps. Some children will be able to do this mentally and describe how they have done it.

ANSWERS

Question **1** a): 5 – 1 = 4
Question **1** b): 5 – 2 = 3
Question **2** a): 8 + 1 = 9
Question **2** b): 8 + 2 = 10
Question **2** c): 8 – 1 = 7
Question **2** d): 8 – 2 = 6
Question **3** a): 7 + 2 = 9
Question **3** b): 7 – 2 = 5

Unit 4: Subtraction within 10, Lesson 7

Think together

1 Use the number line to work out the subtractions.

0 1 2 3 4 5 6 7 8 9 10

Work out:

a) 5 – 1 = ☐
b) 5 – 2 = ☐

2 Work out the additions and subtractions.

○ ○ 8 ○ ○

a) 8 + 1 = ☐
b) 8 + 2 = ☐
c) 8 – 1 = ☐
d) 8 – 2 = ☐

146

PUPIL TEXTBOOK 1A PAGE 146

Unit 4: Subtraction within 10, Lesson 7

CHALLENGE

3 Work out:

a) 7 + 2 = ☐

+ 2

0 1 2 3 4 5 6 7 8 9 10

b) 7 – 2 = ☐

– 2

0 1 2 3 4 5 6 7 8 9 10

+ 2 skips a number.

What about – 2?

147

→ Practice book 1A p105

PUPIL TEXTBOOK 1A PAGE 147

Practice

WAYS OF WORKING Independent thinking

IN FOCUS Questions **1** to **3** provide children with complete number lines to support them in solving a variety of additions and subtractions. Questions **5** and **6** only show abstract calculations.

STRENGTHEN From question **4** onwards, provide children with completed number lines if they need them.

DEEPEN Ask children to create their own addition or subtraction questions of adding/subtracting 1 and 2 which all have the same answer. For example, how many questions can they write with an answer of 4?

WAYS OF WORKING Question **4** asks children to solve additions and subtractions with numbers removed from the number line. This moves children towards mental calculations.

ASSESSMENT CHECKPOINT Children can confidently add or subtract 1 or 2 and show this on a number line. Although children should be moving towards mental calculation, it is not essential at this stage.

ANSWERS Answers for the **Practice** part of the lesson can be found in the *Power Maths* online subscription.

Reflect

WAYS OF WORKING Pair work

IN FOCUS By using the same numbers, this question checks that children are secure with the difference between adding 1 or 2 and subtracting 1 or 2.

ASSESSMENT CHECKPOINT Children can add or subtract 1 or 2 with or without a number line.

ANSWERS Answers for the **Reflect** part of the lesson can be found in the *Power Maths* online subscription.

After the lesson

- Are children confident with the difference between adding 1 or 2 and subtracting 1 or 2?
- Are children able to show the correct number of jumps on a number line and start in the appropriate place?
- Are children counting accurately? Can they describe how to solve problems like this mentally?

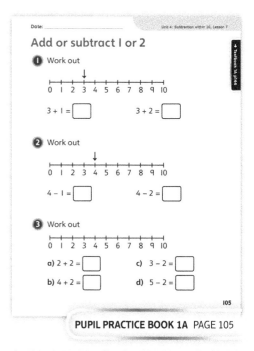

PUPIL PRACTICE BOOK 1A PAGE 105

PUPIL PRACTICE BOOK 1A PAGE 106

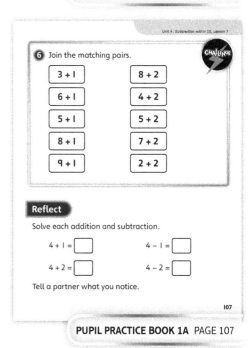

PUPIL PRACTICE BOOK 1A PAGE 107

Solve word problems – addition and subtraction

Learning focus

In this lesson, children will answer a range of addition, subtraction and 'finding the difference' questions. They will make up their own word problems based on a pictorial prompt.

Before you teach

- Are all children familiar with the language surrounding addition and subtraction?
- How could you introduce using a bar model alongside concrete resources?
- How could you scaffold children's ideas, if they are struggling to see which operation needs to be used in a given context?

NATIONAL CURRICULUM LINKS

Year 1 Number – addition and subtraction

Solve one-step problems that involve addition and subtraction, using concrete objects and pictorial representations, and missing number problems such as $7 = \boxed{} - 9$.

ASSESSING MASTERY

Children can read different contexts and problems, identifying the numbers with which they need to work (including the whole, parts and the whole or part that is unknown) and what operations should be used.

COMMON MISCONCEPTIONS

Children may have difficulty interpreting the words in a problem that suggest which operation they need to use. Ask:
- *What parts of the problem do you know? What is the problem asking you to find out?*

STRENGTHENING UNDERSTANDING

Strengthen understanding, for those who struggle to see what calculation to do based on a context, by making a display of common words or phrases that come up in addition and subtraction word problems. For example, write words such as 'total' and 'altogether' next to an addition sign, and phrases such as 'how many are left' next to a subtraction sign.

GOING DEEPER

Ask children to use cubes to solve a problem and then deepen understanding by asking them to represent the cubes as a bar model. (This can be done by making rods of cubes to represent different parts of the problem and drawing around them.) Children should then label the parts of the bar model with the numbers they represent. If this is too abstract, children could draw around each individual cube to create a pictogram of the problem.

KEY LANGUAGE

In lesson: add, total, altogether, subtract, take away, how many are left?

Other language to be used by the teacher: difference, compare, bar model, representation

STRUCTURES AND REPRESENTATIONS

Number line, part-whole model, bar model

RESOURCES

Mandatory: blank number lines, blank part-whole models, multilink cubes, counters

Optional: craft paper, marker pens

 In the eTextbook of this lesson, you will find interactive links to a selection of teaching tools.

Quick recap 🔁

Play subtraction bingo. Ask children to pick four numbers between 0 and 10. Write some subtraction calculations that have answers between 0 and 10. If children have the answer, they cross it off.

Discover

Solve word problems – addition and subtraction

Discover

WAYS OF WORKING Pair work

ASK

- Question ① a): *What numbers do you need in order to work out how many there are in total?*
- Question ① b): *If two apples are sold, are they being added or taken away? From what number are they being taken away?*

IN FOCUS Children must first identify each component in the picture (how many apples there are in the basket and how many there are loose on the table) before identifying the operation needed to solve the problem.

PRACTICAL TIPS Use cubes or counters to replicate the scenario.

ANSWERS

Question ① a): There are 7 apples in total.

Question ① b): There are 5 apples left.

① **a)** How many pieces of fruit are there on the market stall?

b) 2 apples are sold.

How many pieces of fruit are left?

148

PUPIL TEXTBOOK 1A PAGE 148

Share

WAYS OF WORKING Whole class teacher led

ASK

- Question ① a): *How has Astrid used cubes to help?*
- Question ① a): *What has Astrid done with the cubes? How do you think she worked out the total? Are there other ways you can work it out?*
- Question ① b): *Why are two cubes and two apples crossed out?*
- After question ① b): *Why are the arrows on the number lines jumping back?*

IN FOCUS In this part of the lesson, children are prompted to use physical resources to help them work out the problem. Children should start to understand when to use addition and when to use subtraction. Focus on words such as 'total' and 'how many are left?' to help give children guidance. They should write the number sentences alongside the physical representations and use the methods they have learnt to answer the questions. Encourage children to use a variety of different methods to work out the answer and show that, which ever method they use, they will get the same answer.

DEEPEN Ask children to put the numbers from each problem into a part-whole model. Ensure they are clear in what they are trying to find out in each case: in question ① a), the whole is unknown; in ① b), a part is unknown.

Share

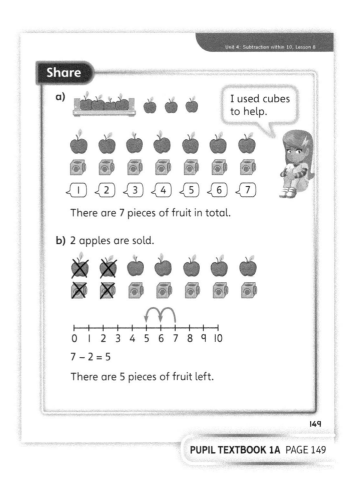

PUPIL TEXTBOOK 1A PAGE 149

187

Think together

Unit 4: Subtraction within 10, Lesson 8

Think together

WAYS OF WORKING Whole class teacher led (I do, We do, You do)

ASK

- Question **①**: *What does each cube represent? How many cubes are out of the sack? How many are in the sack? How can you find the total number of cubes?*
- Question **②**: *Does 8 represent the whole or the part?*
- Question **②**: *How do you represent 5 sweets being eaten? What operation is that?*
- Question **③**: *How can you use cubes to show this problem and what do you do with them?*

IN FOCUS In question **①**, children use concrete objects to represents the presents. They use the method of count on or knowledge of number bonds to find the answer. Some children may need to count all the cubes. In question **②**, children need to realise that this time they have the whole and they are subtracting. They may use a count back or cross out method to find the answer. Discuss which might be most appropriate.

In question **②**, ask children to draw around a rod of 8 cubes, to represent how many sweets there are. Ask them where the 3 is, and what that would look like on this bar. (They could create rods of 5 and 3 cubes to help them.) This will encourage children to think about how bars could represent subtraction.

STRENGTHEN Throughout this **Think together**, ask children if they know the whole or if they need to work out the whole. Focus in detail on the language and what the different numbers represent.

DEEPEN Ask children to deepen their understanding by solving the same problem in different ways and then by making up their own word problem for a given addition or subtraction (e.g. 3 + 4 = 7). What could the question be?

ASSESSMENT CHECKPOINT In question **③**, read what Dexter says about counting the cubes carefully and listen for whether children start counting from 6, from 4 or from 0. Assess whether they are aware that the most efficient method is to count on from 6, or whether they count on from 4 because that is the first group of apples. Children may alternatively be able to recognise this as a number bond to 10 and not need to calculate at all.

ANSWERS

Question **①**: 5 + 4 = 9
There are 9 presents altogether.

Question **②**: 8 − 5 = 3
There are 3 sweets left.

Question **③**: 4 + 6 = 10
There are 10 apples in total.

PUPIL TEXTBOOK 1A PAGE 150

PUPIL TEXTBOOK 1A PAGE 151

Practice

WAYS OF WORKING Independent thinking

IN FOCUS In the first three questions, children are given the addition or subtraction structure to answer the word problems.

STRENGTHEN In question ④, discuss with children how they can tell which sets of cubes show subtraction and which show addition. Children can often struggle with solving word problems. Support children by reading the question as a class. Ask children if they know if they have the whole or just one of the parts. How do they know? Can they represent the situation in a part-whole model? Which number is missing? Help children with words such as 'total' and 'left over' to help them decide whether to add or subtract.

DEEPEN In question ⑤, numbers have been represented using different shapes. Deepen understanding by challenging children to use the same shapes, representing the same numbers, to create their own calculations. Children may find it helps to write the calculations using the numbers first. If children feel secure, they could also assign other numbers to new different shapes, and create more number facts.

THINK DIFFERENTLY In question ④, children are asked to identify which set of cubes matches each number sentence and write in the answers.

ASSESSMENT CHECKPOINT Check that children are able to use the correct operation and put the numbers in the correct place. Check that the answers to the problems match the operation. In question ⑤, assess what strategies children use to work out what shape represents what number. They may recognise that their best starting point is that two matching squares must mean the parts are the same, and that only equal numbers can make 10. Alternatively, they may find this problem too abstract and need further scaffolding.

ANSWERS Answers for the **Practice** part of the lesson can be found in the *Power Maths* online subscription.

Reflect

WAYS OF WORKING Pair work

IN FOCUS Within the context given, children must write their own maths question to match the picture.

ASSESSMENT CHECKPOINT Assess whether children are able to explain what their number sentence is trying to find out, and what each number means with regards to the picture. Check that they can also construct a correct calculation to match their question.

ANSWERS Answers for the **Reflect** part of the lesson can be found in the *Power Maths* online subscription.

After the lesson

- Were children able to recognise the correct operations for the different contexts given?
- What sort of word problems were children able to create independently? What does this tell you about their understanding?

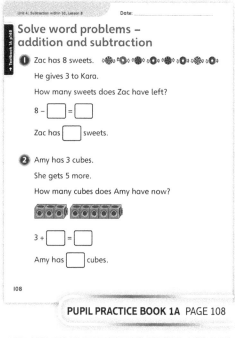

PUPIL PRACTICE BOOK 1A PAGE 108

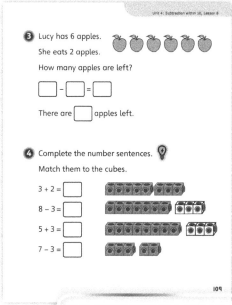

PUPIL PRACTICE BOOK 1A PAGE 109

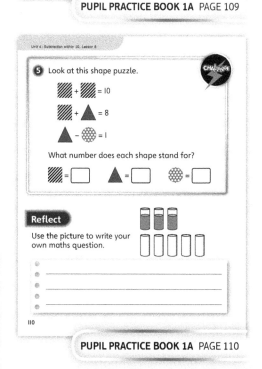

PUPIL PRACTICE BOOK 1A PAGE 110

End of unit check

Don't forget the unit assessment grid in your *Power Maths* online subscription.

WAYS OF WORKING Group work teacher led for questions ❶ to ❺

IN FOCUS

Question ❸ requires children to correctly interpret the part-whole model and understand number sentences with the = sign in different places.

Question ❹ requires children to interpret what subtraction is being shown on the number line. They should relate:
- the starting point shown on the number line (8) as the first number in the number sentence
- the number of jumps back (3) as the number being taken away in the number sentence
- the number they finish on (5) as the answer to the subtraction sentence.

Question ❺ requires children to use a number line to work out the subtraction. They need to decide where the start number is and how many jumps back they need to make.

Think!

WAYS OF WORKING Pair work or small groups

IN FOCUS

- This question highlights the misconception that subtraction is commutative: the numbers in all the facts match the part-whole model but one of them has the whole and one of the parts in the wrong order, and is therefore incorrect. An extra layer of thought is required as the number sentences have the = sign in different places.
- Draw children's attention to the words at the bottom of the **My journal** page. Encourage them to match them to the part-whole model and facts.
- Encourage children to think through or discuss the meaning of the − 6 in 3 − 6 = 3 before writing their answer in **My journal**. It means 'take away six': children should see that this does not match the part-whole model as 6 is the whole.

ANSWERS AND COMMENTARY To show mastery, children can come up with their own number story to match the part-whole model where the parts are the same. In their story, the parts will represent separate and distinct objects but have the same value, for example 3 birds on the branch, 3 flying away or 3 children standing, 3 children sitting.

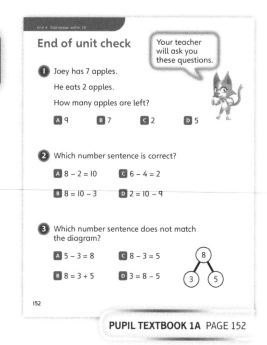

PUPIL TEXTBOOK 1A PAGE 152

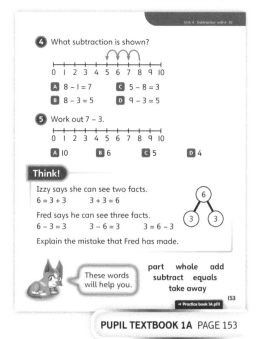

PUPIL TEXTBOOK 1A PAGE 153

Q	A	WRONG ANSWERS AND MISCONCEPTIONS	STRENGTHENING UNDERSTANDING
1	D	A indicates that the child has added 7 and 2 instead of subtracting.	Children may misinterpret these questions as addition, since they are more familiar with this operation. Use concrete resources such as cubes to model taking away from a whole amount, or stacks of cubes to compare and find the difference.
2	C	A suggests that the child has recognised all the numbers needed for a number bond to 10 and not looked at the operation.	
3	A	D indicates a lack of fluency, as it suggests that the child thinks it must be wrong because the part comes first in the number sentence.	
4	B	C suggests that children do not have a solid understanding of the direction of jumps on a number line for subtraction.	
5	D	A suggests that the child has added the numbers together instead of subtracting them.	

My journal

WAYS OF WORKING Independent thinking

ANSWERS AND COMMENTARY

The mistake is 3 – 6 = 3 because this calculation is incorrect and does not match the part-whole model.

When first looking at Fred's statements, some children may immediately point at 3 = 6 – 3 as being incorrect because of the position of the = sign. Encourage children to work out each statement and articulate which numbers are the parts and which is the whole. Can children represent each number sentence using cubes to help them articulate which one is correct? If children struggle to explain, show them the subtraction generalisations 'whole – part = part' and 'part = whole – part' and ask which statement does not match.

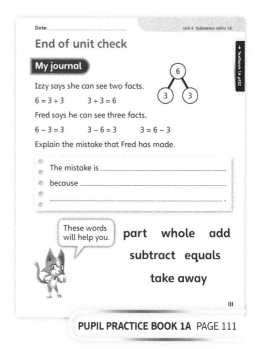

PUPIL PRACTICE BOOK 1A PAGE 111

Power check

WAYS OF WORKING Independent thinking

ASK

- Why do you feel like that about subtraction?
- What helps you work out subtraction questions?

Power puzzle

WAYS OF WORKING Pair work or small groups

IN FOCUS This **Power puzzle** incorporates multiple elements of understanding from this unit and requires children to work strategically. Can children use trial and error, gradually refining their answer to fit? Using real number cards will reinforce the restriction that each number can only be used once.

ANSWERS AND COMMENTARY If children can correctly fill in the **Power puzzle**, they have demonstrated good understanding of multiple operations and signs. Ask them if they can find as many different ways as possible to make each statement true.

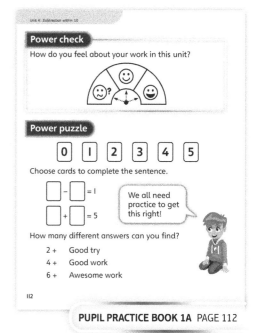

PUPIL PRACTICE BOOK 1A PAGE 112

After the unit

- Were children more confident calculating addition than subtraction number sentences?
- Could children spot the deliberate mistakes? Could children identify if they make the same mistakes in their own work?

Strengthen and **Deepen** activities for this unit can be found in the *Power Maths* online subscription.

Unit 5
2D and 3D shapes

Mastery Expert tip! 'Providing children with 2D and 3D shapes that they could handle and explore not only helps to strengthen understanding but also enables them to think carefully about how to describe the features of different shapes and identify similarities and differences.'

Don't forget to watch the Unit 5 video!

WHY THIS UNIT IS IMPORTANT

This unit introduces children to 2D and 3D shapes and their properties. Children will learn to name the different shapes and identify the features that determine how they are classified. By exploring the similarities and differences, children will make the distinction between 2D and 3D shapes.

Throughout the unit, shapes are presented in different orientations. This helps children focus on the specific mathematical properties of the shapes and secures their understanding of classifying 2D and 3D shapes. They will also learn to identify individual shapes within composite shapes (where several shapes are joined together) and explore the relationship between 2D and 3D shapes.

Children will then begin to explore sequences using shapes and to identify patterns. Children will apply these skills when exploring patterns and sequences in numbers in future units.

WHERE THIS UNIT FITS

→ Unit 4: Subtraction within 10
→ **Unit 5: 2D and 3D shapes**
→ Unit 6: Numbers to 20

This unit builds on the work that children have done on sorting objects. It draws on their skills of identifying similarities and differences and making direct comparisons, and develops their skill of identifying patterns and sequences in shapes. Unit 6 will focus on numbers to 20.

Before they start this unit, it is expected that children:
- know the names of basic 2D and 3D shapes
- understand that shapes are classified based on specific properties
- know that shapes can be sorted using different criteria.

ASSESSING MASTERY

Children who have mastered this unit will be able to identify and describe the key properties of 2D and 3D shapes, using the correct mathematical terminology. They will be able to ignore non-significant differences such as colour, size and orientation in order to classify shapes.

As their shape recognition becomes more secure, they will be able to identify, describe and continue repeating patterns made of shapes.

COMMON MISCONCEPTIONS	STRENGTHENING UNDERSTANDING	GOING DEEPER
Children may apply the names of 2D shapes to 3D shapes or vice versa.	Give children concrete representations of the shapes and encourage them to talk about their similarities and differences. Provide children with labels to match with the shapes.	Ask children to find what different shapes have in common. They could explore combining 2D or 3D shapes to create a different shape.
Children may fail to recognise a shape when its orientation changes and may focus on superficial differences such as colour or size.	Ask children to sort a variety of different shapes using their properties. As they do so, discuss the names of the shapes and how we know what the shapes are called.	Ask children to identify shapes that are only partially revealed or by touch alone. Ask children to justify their decisions.

Unit 5: 2D and 3D shapes

UNIT STARTER PAGES

These pages will help you identify children's prior knowledge of the names and properties of 2D and 3D shapes. Focus on the vocabulary introduced by Flo: ask children what each term means and assess the accuracy and confidence of their answers.

STRUCTURES AND REPRESENTATIONS

It is important that children have a range of 2D and 3D shapes to manipulate and explore. These should include: cubes, cuboids, spheres, cylinders, pyramids, cones, circles, triangles, squares, rectangles.

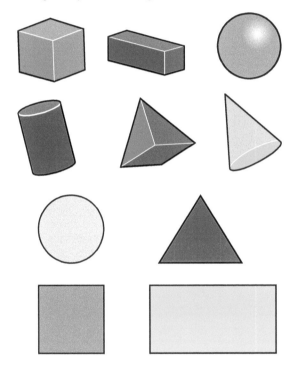

KEY LANGUAGE

It is important that children describe shapes using the correct mathematical terminology.

→ 2D shape, 3D shape

→ cube, cuboid, sphere, cylinder, pyramid, cone

→ circle, triangle, square, rectangle

→ sides, edges, faces, corners

→ pattern, repeated

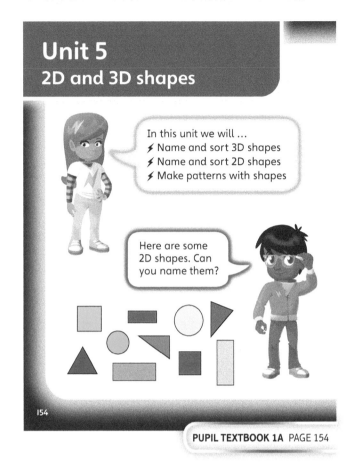

PUPIL TEXTBOOK 1A PAGE 154

PUPIL TEXTBOOK 1A PAGE 155

Recognise and name 3D shapes

Learning focus

In this lesson, children will learn to name and identify 3D shapes. They will compare shapes, identifying their similarities and differences.

Before you teach

- What resources will you need to support children's understanding of 3D shapes?
- How could you display key terminology for children to refer to?
- What experience of 3D shapes do children have?

NATIONAL CURRICULUM LINKS

Year 1 Geometry – properties of shape

Recognise and name common 2D and 3D shapes, including: 3D shapes [for example, cuboids (including cubes), pyramids and spheres].

ASSESSING MASTERY

Children can identify the common properties of the same type of 3D shape even when they are visually different (for example, different size, colour, orientation or dimensions). Children can use the correct terminology for the properties in order to justify their reasoning.

COMMON MISCONCEPTIONS

Children may misname 3D shapes as 2D shapes. They may fail to name a shape correctly when presented with an unfamiliar orientation. Give children concrete 3D shapes to explore and ask:
- *What shape is this?*

They may name a shape as a familiar object, for example, as a 'ball', 'dice' or 'box'. To challenge this misconception, ask:
- *What type of shape is a dice?* Reinforce your question by showing children a multi-sided dice.

STRENGTHENING UNDERSTANDING

To strengthen understanding, show children a selection of concrete 3D shapes and everyday items that correspond to those shapes. Ask children to pair up or sort the shapes using sorting hoops. Ask: *How are these shapes the same or different?* Ask children to rotate the shapes into different orientations. Ask: *Has the shape changed?*

GOING DEEPER

Show children a cube, cuboid and cylinder made out of modelling material. Tell children these can all be grouped together. Ask: *What do the shapes have in common? What will happen to the end face of each shape if I cut them in half?* Children can predict what will happen and then find out by trying it themselves. The end face does not change shape or size and is the same for all three shapes: this defines a prism. Ask: *Would this be true for any of the other shapes? Why?*

KEY LANGUAGE

In lesson: 3D shapes, **cube**, **cuboid**, **sphere**, **pyramid**, **cylinder**, shape, pair, same, different

Other language to be used by the teacher: similarities, differences, properties, face, edge

STRUCTURES AND REPRESENTATIONS

Pictorial representations of cubes, cuboids, spheres, cylinders, pyramids, cones

RESOURCES

Mandatory: concrete 3D shapes including a cube, cuboid, sphere, cylinder, pyramid

Optional: modelling material to model 3D shapes, sorting hoops, an opaque bag, everyday items relating to the 3D shapes (for example, a golf ball, a cereal box, an empty sweet tube, dice)

In the eTextbook of this lesson, you will find interactive links to a selection of teaching tools.

Quick recap

What 3D shapes do children recognise? Show some 3D shapes and ask children to name each one. Introduce any names that children do not yet know.

Discover

Recognise and name 3D shapes

Discover

These are all examples of 3D shapes.

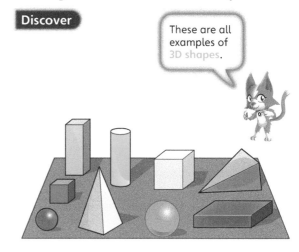

WAYS OF WORKING Pair work

ASK

• Question ① a): *Why have you grouped those shapes together? How are they the same or different? How do you know which shapes are cubes? Can you pair them differently?*

IN FOCUS Question ① a) highlights that shapes are alike based on certain properties and that size or colour do not distinguish alike shapes. Determine if children are using the correct terminology and names. Are children basing their decisions on key properties? What do they already know? What misconceptions do they have?

PRACTICAL TIPS Provide real-life 3D shapes (balls, cube or cuboid-shaped packaging, etc.) to recreate the scenario in the classroom. Again, ask children to put shapes with the same name together.

① a) Put shapes with the same name together.

b) What is the shape without a pair?

156

ANSWERS

Question ① a):

   cube   cuboid   sphere   pyramid

Question ① b): The cylinder does not have a pair.

PUPIL TEXTBOOK 1A PAGE 156

Share

WAYS OF WORKING Whole class teacher led

ASK

• Question ① a): *What properties did you look at in order to pair your shapes? What properties did you ignore? Did you find any shapes difficult to pair? Could you have paired them differently? Are cubes and cuboids the same?* (Cubes are a special type of cuboid. A cuboid is made up of six rectangles placed at right angles to one another. A cuboid that has six square faces is given the special name 'cube'.)

IN FOCUS Question ① a): Astrid's statement reinforces that some properties, such as colour and orientation, are not important when naming 3D shapes. You could follow this up in class by comparing concrete 3D shapes.

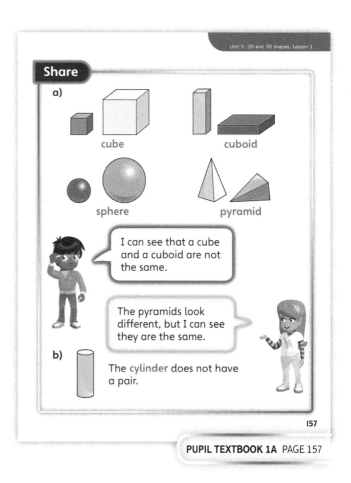

PUPIL TEXTBOOK 1A PAGE 157

195

Think together

WAYS OF WORKING Whole class teacher led (I do, We do, You do)

ASK

- Question **1**: *How do you know that the other shapes are not cubes?*
- Question **2** a): *What do all pyramids have in common?*
- Question **2** b): *What makes the sphere different from the ovoid?*
- Question **3**: *How do you know what the shapes are called?*

IN FOCUS Question **3** prompts children to look beyond the aesthetic properties of an object and focus on the properties of the shapes. It distinguishes between the name of the object and the more abstract name of the shape. It prompts children to realise that 3D shapes can be found in their everyday environment.

STRENGTHEN Having concrete 3D shapes for children to hold and manipulate can help them to identify the shapes in the pictures. Encourage children to rotate them so that they match the orientations in the pictures. Ask: *Does turning the shape change the shape?*

DEEPEN Ask children to find two different types of shapes that have similarities and to explain how they are similar. Ask: *How many different ways could you group the shapes?*

ASSESSMENT CHECKPOINT Questions **1** and **2** a) can help you determine whether children can see beyond the orientation and colour of the shape.

Children's responses to question **3** will determine whether they can focus on the properties of shapes. Ask children to describe the properties of the shapes that they identify. Are they counting and describing faces? Can they describe the differences between different shapes?

ANSWERS

Question **1**: The first, fourth and fifth shapes are cubes.

Question **2** a): There are 2 pyramids.

Question **2** b): 2 of the shapes are spheres.

Question **3**:

      sphere   cylinder   cuboid   cube

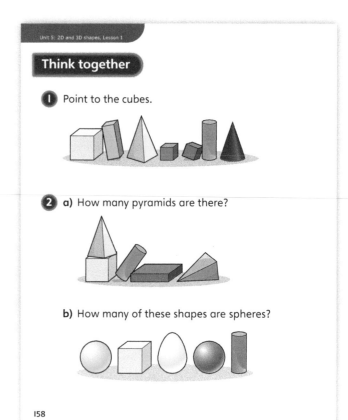

Think together

1 Point to the cubes.

2 a) How many pyramids are there?

   b) How many of these shapes are spheres?

158

PUPIL TEXTBOOK 1A PAGE 158

Think together

3 What is the mathematical name of each of these shapes?

CHALLENGE

The first one is just a football. Is that the name of the shape?

I think the shape of each of these objects has a special name.

159

→ Practice book 1A p113

PUPIL TEXTBOOK 1A PAGE 159

Practice

WAYS OF WORKING Pair work

IN FOCUS Question ① asks children to identify 3D shapes from pictorial representations. Question ② asks children to identify the odd one out from pictorial representations of 3D shapes.

STRENGTHEN Have some of the objects pictured in question ③ for children to hold, such as the dice and the golf ball. This will enable them to make comparisons between the concrete shapes and the pictorial representations.

DEEPEN Put shapes in an opaque bag so that the shapes cannot be seen. Ask a child to put their hand into the bag and hold a shape without taking it out. Can they identify it by touch alone? Ask: *How do you know what it is?* They could then describe the shape they are holding for another child to identify.

THINK DIFFERENTLY Question ③ looks at everyday objects, encouraging children to focus on the properties of shapes in order to identify them. One object is a compound shape, which is a shape consisting of two or more basic shapes (in this case, a sphere and a cylinder).

ASSESSMENT CHECKPOINT Questions ① and ② assess whether children can identify the shapes, regardless of size and orientation.

Question ③ determines whether children can focus solely on the properties of the shapes in order to identify the shapes.

Question ④ prompts children to draw on their knowledge in order to identify incomplete shapes.

ANSWERS Answers for the **Practice** part of the lesson can be found in the *Power Maths* online subscription.

Reflect

WAYS OF WORKING Independent thinking

IN FOCUS The **Reflect** part of the lesson prompts children to relate the properties of 3D shapes to their everyday experiences. They need to think of the objects by their properties of shape, ignoring other features such as the object's name, size, colour, pattern and purpose.

Refer to what Sparks is asking. Can children use their broader experiences to suggest where pyramids are found?

ASSESSMENT CHECKPOINT Assess whether children are able to identify objects that are the same shapes as those pictured. Check whether they can think of an example for all four shapes (or all five shapes if you include Sparks' question).

ANSWERS Answers for the **Reflect** part of the lesson can be found in the *Power Maths* online subscription.

After the lesson ⏸

- Were children confident in identifying the 3D shapes, regardless of their colour, size and orientation?
- Were children able to stop relying on the concrete 3D shapes or did some children still need concrete 3D shapes to support their reasoning?
- Were children starting to use mathematical terminology in order to describe 3D shapes and their properties?

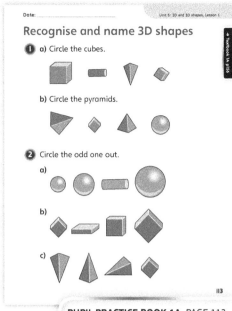

PUPIL PRACTICE BOOK 1A PAGE 113

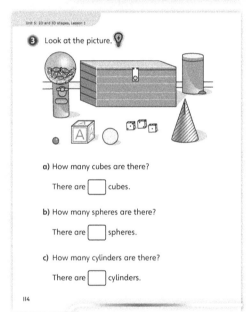

PUPIL PRACTICE BOOK 1A PAGE 114

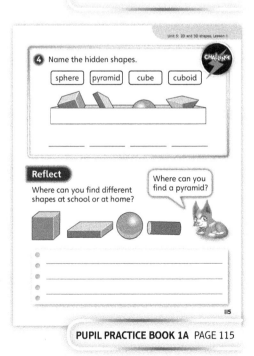

PUPIL PRACTICE BOOK 1A PAGE 115

Sort 3D shapes

Learning focus

In this lesson, children will build on their ability to name and describe 3D shapes in more challenging arrangements. They will also be introduced to the cone.

Before you teach ⏸

- How confident are children in identifying and naming the shapes from Lesson 1?
- How will you support children who find it difficult to identify shapes when they are connected to other shapes?
- How will you question and support children in order to develop the use of correct mathematical vocabulary when they describe the properties of shapes?

NATIONAL CURRICULUM LINKS

Year 1 Geometry – properties of shape

Recognise and name common 2D and 3D shapes, including: 3D shapes [for example, cuboids (including cubes), pyramids and spheres].

ASSESSING MASTERY

Children can use correct terminology to describe the properties of a range of 3D shapes. Children can make careful comparisons and identify both differences and similarities between different 3D shapes.

COMMON MISCONCEPTIONS

Children may believe that a curved surface is a face, such as on a sphere or cone. Remind them that a face has to be flat. Ask:
- *How many faces are there? How many curved surfaces are there? What is the difference between a face and a curved surface?*

Children may see cubes as a separate shape to a cuboid. Give children some 3D shapes or images of 3D shapes, including multiple cuboids (some of which are cubes). Ask:
- *How many cuboids are there altogether?*

STRENGTHENING UNDERSTANDING

The use of concrete 3D shapes can help children to identify pictorial representations of the same shapes. As you work through each learning section, encourage children to match the arrangements in the pictures with the concrete 3D shapes and ask them to identify which shapes they have used. You can offer further support by labelling the concrete 3D shapes in the classroom, then cover the labels as children become more confident.

GOING DEEPER

Ask children to find similarities between two different shapes. Ask: *What are the similarities between a cone and a pyramid? How are a cylinder and a cuboid alike?*

KEY LANGUAGE

In lesson: 3D, cube, cuboid, sphere, cylinder, pyramid, **cone**, shape, how many, same

Other language to be used by the teacher: similarities, differences, properties, edges, faces, square, triangular, circular, curved, curved surface, flat

STRUCTURES AND REPRESENTATIONS

Pictorial representations of cubes, cuboids, spheres, cylinders, pyramids, cones

RESOURCES

Mandatory: concrete 3D shapes including a cube, cuboid, sphere, cylinder, pyramid, cone

Optional: 3D shape name labels, everyday items relating to the 3D shapes (for example, a golf ball, a cereal box, an empty sweet tube, dice)

 In the eTextbook of this lesson, you will find interactive links to a selection of teaching tools.

Quick recap 🔁

Give children some 3D shapes or images of 3D shapes. Ask them to sort the objects into groups. Can they describe how they have sorted them?

Discover

Sort 3D shapes

Discover

WAYS OF WORKING Pair work

ASK

- Question ① a): *Which shapes can you see? How many of each shape can you see? What are the similarities and differences between the rockets?*
- Question ① b): *Can you name and describe each shape?*

IN FOCUS Question ① a) requires children to recognise similarities and differences between the two rockets by identifying and describing shapes in different orientations.

PRACTICAL TIPS Give children concrete 3D shapes and ask them to build the two rockets.

ANSWERS

Question ① a): Rocket 1 has broken.

Question ① b): The sphere and the cone were not used.

❶ **a)** Which rocket has broken?

  b) Which of these shapes were not used?

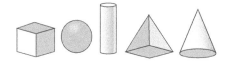

160

PUPIL TEXTBOOK 1A PAGE 160

Share

WAYS OF WORKING Whole class teacher led

ASK

- Question ① a): *Can you create the two rockets for yourself using shapes?*
- Question ① a): *How do you know that rocket 1 is the rocket that broke?*
- Question ① b): *Why do you think that the sphere and the cone were not used? Which shapes are easy or difficult to tell apart?*

IN FOCUS Question ① a) models the sequence of shapes used to create the rockets, supporting children in separating the rockets out into the individual component shapes.

The pictures of the sphere and the cone in question ① b) give scaffolding for children's reasoning and provide the proof for the answer.

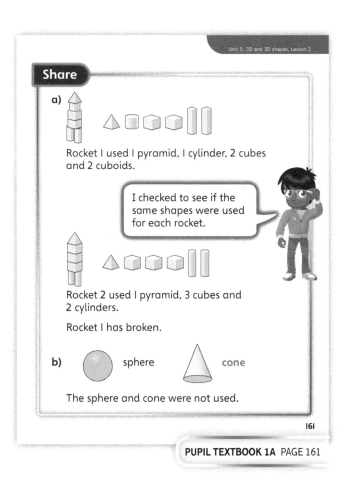

PUPIL TEXTBOOK 1A PAGE 161

199

Think together

Think together

WAYS OF WORKING Whole class teacher led (I do, We do, You do)

ASK

· *How could you make it easier to identify the shapes?*
What are the similarities between the different shapes?
What properties of the shapes do you look at first?
What is it about the pictures that makes identifying the shapes easier?

IN FOCUS Question ❸ prompts discussion about cubes being a type of cuboid. It encourages children to identify the properties that they have in common. Can children identify that squares are a type of rectangle?

STRENGTHEN Allow children access to concrete 3D shapes. They can use these shapes to replicate the pictures in this part of the lesson, then dismantle the compound shapes they have created to sort and identify the individual shapes that they used.

DEEPEN Can children identify that all cuboids have six rectangular faces, that some cuboids have only square faces, that some cuboids have only oblong faces and that some cuboids have a mixture of square and oblong faces? To deepen understanding, refer to Ash's and Flo's questions. Ask: *What can you say is true of all cuboids? What can you say is true for some but not all cuboids?*

ASSESSMENT CHECKPOINT Assess whether children can recognise cubes as cuboids and identify their common properties.

ANSWERS

Question ❶ a): Cube and pyramid.

Question ❶ b): Cuboid and cone.

Question ❶ c): Cuboid, cube and cylinder.

Question ❷ a): There are 3 cubes.

Question ❷ b): There are 2 cylinders.

Question ❷ c): There are 0 spheres.

Question ❷ d): There are 3 pyramids.

Question ❸: There are 5 cuboids, 3 of which are also cubes.

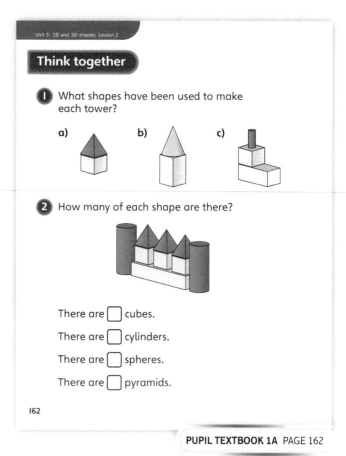

PUPIL TEXTBOOK 1A PAGE 162

PUPIL TEXTBOOK 1A PAGE 163

Practice

WAYS OF WORKING Pair work

IN FOCUS Question **3** requires children to identify the individual shapes within compound shapes, requiring careful observation and application of knowledge.

STRENGTHEN Have concrete 3D shapes so that children can replicate the pictures in this part of the lesson and identify the shapes that they use. Labelling the shapes can provide further support if children still struggle to recall the names of the shapes.

DEEPEN Describe a shape, one property at a time. After describing each property, ask: *Which shape do you think it could be?* How many properties do you need to describe before children can identify exactly which shape it is? Children could then question you about a shape that you are thinking about, in order to identify the shape in your mind.

THINK DIFFERENTLY Question **4** asks children to match groups of every day objects to their corresponding 3D shapes.

ASSESSMENT CHECKPOINT Assess whether children can recognise cubes as cuboids. Prompt them to count and describe the faces of the shapes.

Questions **4** and **5** will help you determine whether children are able to focus on the properties of the shapes and ignore the context and aesthetic properties of the shapes.

ANSWERS Answers for the **Practice** part of the lesson can be found in the *Power Maths* online subscription.

Reflect

WAYS OF WORKING Independent thinking

IN FOCUS The **Reflect** part of the lesson requires children to use correct mathematical terminology and to recall what they have learnt without visual prompts.

ASSESSMENT CHECKPOINT Assess whether children use the correct terminology and can correctly spell the names of the shapes. Refer to Astrid's comment. Check whether children can find an example of each shape in the classroom and label them.

ANSWERS Answers for the **Reflect** part of the lesson can be found in the *Power Maths* online subscription.

After the lesson ⏸

- Were children able to identify 3D shapes in their environment?
- How carefully were children considering the properties of the 3D shapes when they compared them, particularly with cubes and other cuboids?
- Were children using the correct terminology to name and describe different 3D shapes?

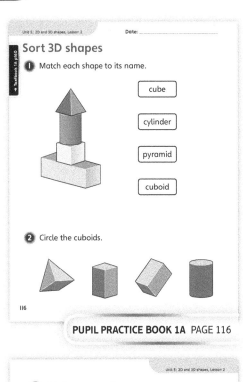

PUPIL PRACTICE BOOK 1A PAGE 116

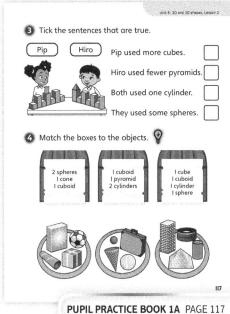

PUPIL PRACTICE BOOK 1A PAGE 117

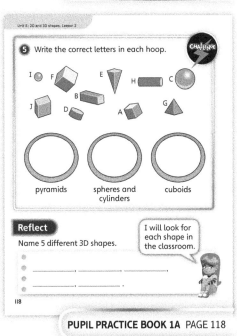

PUPIL PRACTICE BOOK 1A PAGE 118

Recognise and name 2D shapes

Learning focus

In this lesson, children will learn to identify and name 2D shapes: circles, triangles, rectangles and squares.

Before you teach

- How will you explain the difference between 2D and 3D shapes?
- How will you encourage children to use correct mathematical vocabulary to describe 2D shapes?
- How will you draw on children's knowledge from the previous lessons on 3D shapes?

NATIONAL CURRICULUM LINKS

Year 1 Geometry – properties of shape

Recognise and name common 2D and 3D shapes, including: 2D shapes [for example, rectangles (including squares), circles and triangles].

ASSESSING MASTERY

Children can identify, name and describe a circle, square, rectangle and triangle, regardless of the shapes' colour, orientation and size. Children can identify non-examples of the shapes and explain why they do not meet the criteria of specific shapes.

COMMON MISCONCEPTIONS

Children may fail to identify shapes correctly when the shapes are rotated. Hold up a square and ask:
- *Can you name this shape?* Then rotate the square by 45 degrees. *Now, can you name the shape?*

Children may only recognise equilateral triangles and fail to recognise other types of triangles as triangles. Ensure that representations of all types of triangle are available. Children do not need to know the name of the shapes at this stage, they just need to recognise them all as triangles. Ask:
- *What shape are these? Are they the same shape even though they look different?*

STRENGTHENING UNDERSTANDING

Have a variety of concrete 2D shapes available for children to handle. Ask children to create pictures with them and encourage them to name the shapes that they use.

GOING DEEPER

Encourage children to explore the 2D concrete shapes more deeply by themselves. Ask: *Can you use two or four triangles to create a rectangle? Can you fold a rectangle or square to make a triangle? What different triangles can you make?*

KEY LANGUAGE

In lesson: 2D shapes, triangles, circles, squares, rectangles, spheres, corners

Other language to be used by teacher: sides, curved, straight

STRUCTURES AND REPRESENTATIONS

Pictorial representations of 2D shapes of different sizes, including: squares, rectangles, triangles, circles, parallelograms

RESOURCES

Mandatory: a selection of concrete 2D representations of squares, rectangles, triangles, circles, parallelograms

Optional: sorting hoops

 In the eTextbook of this lesson, you will find interactive links to a selection of teaching tools.

Quick recap

What 2D shapes do children recognise? Show them some cut outs of 2D shapes or images of 2D shapes on the board. Which ones do they know already?

Discover

Recognise and name 2D shapes

WAYS OF WORKING Pair work

ASK

- Question ① a): *How do you know which shapes in the picture are triangles? How do you know which are circles? And rectangles? How is the blue blob similar to a circle?*
- Question ① b): *Did any of the shapes surprise you? Why?*

IN FOCUS Question ① b) requires children to find shapes that are not triangles. Children need to be secure with the properties of a triangle in order to find shapes that do not meet that criteria. The irregular pentagon has a similar appearance to a triangle, so ensure that children look closely at the number of sides and corners on this shape.

PRACTICAL TIPS Give children concrete 2D shapes, including circles, rectangles and triangles, so they can become familiar with their properties.

ANSWERS

Question ① a): There are 2 circles, 3 squares, 3 triangles, 2 rectangles (5 including the squares) and 2 other shapes.

Question ① b): 9 of the shapes are not triangles.

Discover

These are all examples of 2D shapes.

① a) How many of each shape are there?

  b) How many of the shapes are **not** triangles?

164

PUPIL TEXTBOOK 1A PAGE 164

Share

WAYS OF WORKING Whole class teacher led

ASK

- Question ① a): *Does it matter if the shapes in each row are different colours or sizes? Why? What is different about the triangles? Do these shapes remind you of any 3D shapes? Why? What makes a triangle a triangle?*

IN FOCUS Question ① a) requires children to sort shapes into groups, ignoring differences in colour and orientation.

Question ① b) sorts the shapes into two discrete sets: 'triangles' and 'not triangles'. It shows how shapes can be grouped by 'yes' or 'no' criteria. Discuss with children how such a variety of shapes ended up in one set, just because they do not have three straight sides and three corners.

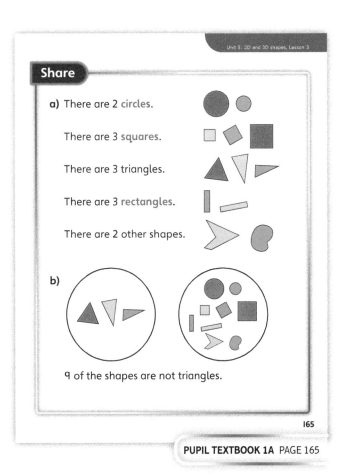

Share

a) There are 2 circles.

  There are 3 squares.

  There are 3 triangles.

  There are 3 rectangles.

  There are 2 other shapes.

b)

  9 of the shapes are not triangles.

165

PUPIL TEXTBOOK 1A PAGE 165

Think together

WAYS OF WORKING Whole class teacher led (I do, We do, You do)

ASK

- For all questions: *How do you know whether a shape is a square, a triangle, a circle or a rectangle?*
- Questions **1** and **2**: *Which shapes did you rule out first? Were there any shapes that you had to think more carefully about?*

IN FOCUS Ask children what the shapes in question **3** have in common (for example, number of sides or number of corners). The question supports children in starting to think about the properties of a rectangle before they formally start using the language of the properties of shape in Year 2. If children say that they all have four corners, respond: *Does that mean anything with four corners is a rectangle?* If children start to talk about squares, refer to the *Deepen* section below. At this stage, children are not expected to know the word 'parallelogram'. They should be able to describe what makes it different from the other shapes in their own language. They may have the misconception that this shape is a rectangle since it has four sides and four corners. In this case, prompt them to think about the corners.

STRENGTHEN Have a selection of concrete 2D shapes available for children to handle. Encourage them to touch the sides and corners as they count them and to physically sort the shapes, perhaps using sorting hoops.

DEEPEN Draw a square and ask children to consider whether this shape would also be an odd one out in the group of shapes from question **3**. Encourage children to think about how the square is different from the other rectangles. Explain that a square is a special kind of rectangle and ask: *What makes it special?* You could also draw a triangle and encourage children to describe the differences between the two shapes, encouraging them to interpret the shapes in their own words.

ASSESSMENT CHECKPOINT Assess whether children are confident in identifying examples and non-examples of a rectangle, square or triangle. Check whether they can begin to describe the features of these shapes to justify their answers.

ANSWERS

Question **1**: There are 3 squares. There are 2 circles.

Question **2**: There are 2 rectangles. There are 3 triangles.

Question **3**: The third shape (the parallelogram) is the odd one out because it is not a rectangle.

Think together

1 Point to the squares and circles.

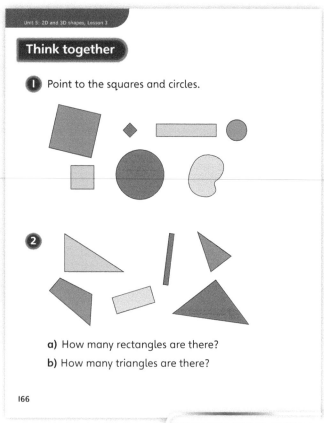

2

a) How many rectangles are there?

b) How many triangles are there?

166

PUPIL TEXTBOOK 1A PAGE 166

3 Point to the odd one out.

CHALLENGE

I'm not sure if they all have 4 corners.

167

→ Practice book 1A p119

PUPIL TEXTBOOK 1A PAGE 167

Practice

WAYS OF WORKING Pair work

IN FOCUS Question ❹ introduces the idea that 2D shapes can be created by combining other 2D shapes. There are two possible answers to this question: a 1 × 6 arrangement or a 2 × 3 arrangement. This question encourages children to think creatively. If children do not find the two possibilities in their pair, ask: *Is that the only way of doing it?*

STRENGTHEN Have a selection of concrete 2D shapes for children to handle. If necessary, help children to label the concrete 2D shapes and talk about their properties. Can children compare the concrete shapes with the pictorial representations in the questions?

DEEPEN Extend question ❹ by asking: *What other 2D shapes can you make from different 2D shapes? Can you create a rectangle by using two triangles? Can you use circles to create other shapes? Why?* While exploring this, children may create shapes that they do not recognise. Reassure them that, even though they do not recognise these shapes, they are still shapes.

THINK DIFFERENTLY In question ❸, children are asked to identify 2D shapes from pictorial representations that are partially hidden.

ASSESSMENT CHECKPOINT Assess whether children can identify and describe a circle, rectangle, triangle and square. Question ❷ should help you check that children can identify non-examples of a shape and explain why.

Question ❺ allows you to assess whether children can identify individual 2D shapes when they are part of a compound shape.

ANSWERS Answers for the **Practice** part of the lesson can be found in the *Power Maths* online subscription.

Reflect

WAYS OF WORKING Independent thinking

IN FOCUS The **Reflect** part of the lesson asks children to identify 2D shapes that are partly obscured. Children need to draw on their knowledge and experience of 2D shapes in order to identify them. To expand their thinking, ask children to describe how they can be sure that the shape is a triangle or a rectangle. This will challenge them to be flexible and creative in their thinking.

ASSESSMENT CHECKPOINT Assess whether children are using the correct mathematical terminology to name and describe 2D shapes.

ANSWERS Answers for the **Reflect** part of the lesson can be found in the *Power Maths* online subscription.

After the lesson ⏸

- Did the use of concrete 2D shapes enable children to understand the pictorial representations?
- Were children able to make links between 2D shapes and 3D shapes?
- Were you able to challenge children's thinking and encourage them to provide justifications when identifying shapes?

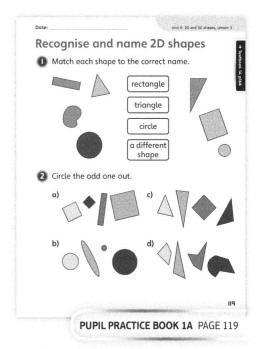

Date: _____

Recognise and name 2D shapes

❶ Match each shape to the correct name.

rectangle
triangle
circle
a different shape

❷ Circle the odd one out.

a)   c)
b)   d)

119

PUPIL PRACTICE BOOK 1A PAGE 119

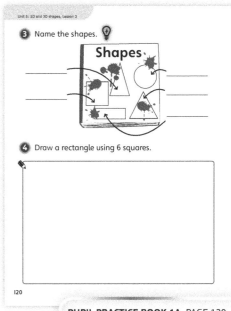

Unit 5: 2D and 3D shapes, Lesson 3

❸ Name the shapes. 💡

Shapes

❹ Draw a rectangle using 6 squares.

120

PUPIL PRACTICE BOOK 1A PAGE 120

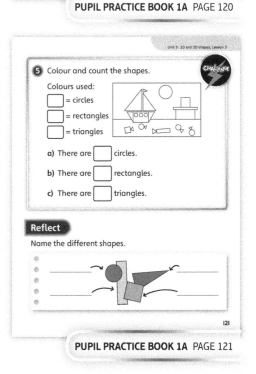

Unit 5: 2D and 3D shapes, Lesson 3

❺ Colour and count the shapes. **CHALLENGE**

Colours used:
☐ = circles
☐ = rectangles
☐ = triangles

a) There are ☐ circles.
b) There are ☐ rectangles.
c) There are ☐ triangles.

Reflect

Name the different shapes.

121

PUPIL PRACTICE BOOK 1A PAGE 121

Sort 2D shapes

Learning focus

In this lesson, children will link their learning from previous lessons in Unit 5, understanding the relationship between 2D and 3D shapes.

Before you teach

- How secure are children in distinguishing between and naming 2D and 3D shapes?
- Are children able to describe the properties of 2D and 3D shapes?
- How will you use concrete 2D and 3D shapes to strengthen children's understanding?

NATIONAL CURRICULUM LINKS

Year 1 Geometry – properties of shape

Recognise and name common 2D and 3D shapes, including: 2D shapes [for example, rectangles (including squares), circles and triangles]; 3D shapes [for example, cuboids (including cubes), pyramids and spheres].

ASSESSING MASTERY

Children can identify and name the faces of 3D shapes. They can envisage how 3D shapes can be used to create 2D shapes and identify which 3D shapes can be used to create 2D shapes.

COMMON MISCONCEPTIONS

Children may still be confused about naming 2D and 3D shapes, for example, by calling a cube a square. Ask:
- *How is a square different to a cube?*
- *Can you find a square on the cube?*

They·may also believe that a face is the same as a curved surface. Ask:
- *How many faces are there on a sphere? How many curved surfaces are there on a sphere?*
- *What is the difference between a face and a curved surface?*
- *What 3D shapes can you remember? What 2D shapes can you remember?*

STRENGTHENING UNDERSTANDING

Have a selection of concrete 3D shapes available and invite children to explore making 2D shapes by printing using paint and concrete 3D shapes. Have a display of vocabulary relating to 2D and 3D shapes (for example, face, edge, side, curved surface) for children to refer to.

GOING DEEPER

Ask children to predict what shapes can be printed using the faces and curved surfaces of the cone, cylinder, hemisphere and sphere. Ask them to justify their predictions and then test these predictions using paint and the concrete 3D shapes.

KEY LANGUAGE

In lesson: 2D shapes, 3D shapes, cube, cone, cuboid, cylinder, sphere, pyramid, **faces**, triangle, square, circle, rectangle, overlap, different

Other language to be used by the teacher: sides, edges, corners

STRUCTURES AND REPRESENTATIONS

Pictorial representations of 3D and 2D shapes, including cubes, cones, cuboids, cylinders, spheres, pyramids, hemispheres, triangles, squares, circles, rectangles

RESOURCES

Mandatory: concrete 3D representations of cubes, cones, cuboids, cylinders, spheres, pyramids, cones, hemispheres

Optional: dry-wipe markers, paint and paper for shape-printing, sorting hoops, 3D shape name labels, display of vocabulary relating to 2D and 3D shapes (for example, face, edge, side, curved surface)

 In the eTextbook of this lesson, you will find interactive links to a selection of teaching tools.

Quick recap 🔾

Give children the names of some 2D and 3D shapes. Ask them to find or draw an object that matches each name.

Discover

Unit 5: 2D and 3D shapes, Lesson 4

Sort 2D shapes

Discover

WAYS OF WORKING Pair work

ASK

- Question ① a): *Which 3D shape has a face that is a square/ rectangle/triangle/circle …?*
- Question ① a): *Which faces of the 3D shapes did Kat not use?*

IN FOCUS Question ① a) encourages children to make links between 2D and 3D shapes. It helps them to isolate a specific property of a shape and look carefully at their similarities and differences.

PRACTICAL TIPS Ask children to recreate the artwork that Kat has printed, using paint and concrete 3D shapes or 3D sponges.

ANSWERS

Question ① a): Kat used the cube for the square head. She used the cone and the cuboid for the body.

Question ① b): Kat printed the arms first because the hands overlap the arms.

❶ a) Which 3D shapes did Kat use to print the head and the body?

b) Did Kat print the arms or the hands first?

168

PUPIL TEXTBOOK 1A PAGE 168

Share

WAYS OF WORKING Whole class teacher led

ASK

- Question ① a): Refer to what Dexter and Flo say by asking: *Could Kat have used another shape to make the head?*
- Question ① a): Refer to what Flo says about cuboids having square faces by asking: *Is this true?*
- Question ① b): *What do you think the first shape Kat printed was?*

IN FOCUS Question ① b) asks children to think about the order in which the picture is printed. This encourages them to use positional language and provide justification for their answers. They may generate more than one possibility. This can prompt discussion that, sometimes in mathematics, there may not be a definitive answer and that answers should be supported with reasoning and justification.

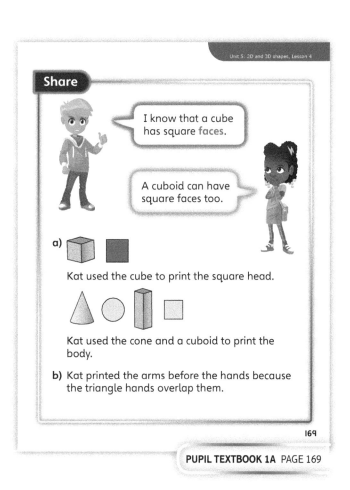

Share

I know that a cube has square faces.

A cuboid can have square faces too.

a) Kat used the cube to print the square head.

Kat used the cone and a cuboid to print the body.

b) Kat printed the arms before the hands because the triangle hands overlap them.

169

PUPIL TEXTBOOK 1A PAGE 169

Think together

Whole class teacher led (I do, We do, You do)

ASK

- Questions **1** and **2**: *What shapes can you use to print a square?*
- Question **1**: *What different rectangles can you can see on the two cuboids?*
- Question **2**: *How many different 2D shapes could you print with the two pyramids? How does this knowledge help you to describe the 3D shapes?*

IN FOCUS Question **3** challenges the misconception that a curved surface is a circle, which often comes from knowing that a circle has one curved side. Discuss the fact that a flat circular face is needed to print a circle. Some children may see that the curved surface of the cone can be rolled, so it is not flat.

STRENGTHEN Have a selection of real-life objects shaped as cuboids and pyramids available for children to handle. In question **1**, children can explore printing or drawing round each face to determine what rectangles these shapes can produce. In question **2**, children can tick each face with a pen on the interactive whiteboard as they count them to help them keep track of which faces they have counted.

DEEPEN In question **3**, ask children what shapes the curved surfaces of a cylinder and a cone would print if rolled. Children can predict, justify their predictions and test what happens.

ASSESSMENT CHECKPOINT Question **2** should determine whether children can count and describe all faces of the pyramids. Can children use their knowledge of 3D shapes to count hidden faces?

Refer to Dexter's statement. Question **3** will expose any insecurities in children's knowledge that a flat face will print a 2D shape. Assess whether children can identify curved surfaces of 3D shapes.

ANSWERS

Question **1** a): The cuboid can print squares and rectangles.

Question **1** b): The cuboid can print 3 different rectangles.

Question **2** a): Both the square-based pyramid and the triangular-based pyramid have 4 triangular faces.

Question **2** b): The square-based pyramid has 1 square face. The triangular-based pyramid does not have any square faces.

Question **3**: The cylinder, the cone and the hemisphere can all print a circle. The sphere cannot print a circle.

Unit 5: 2D and 3D shapes, Lesson 4

Think together

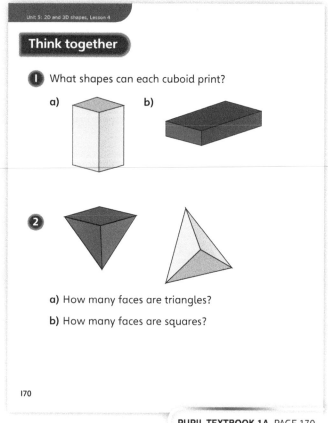

1 What shapes can each cuboid print?

a)   b)

2

a) How many faces are triangles?

b) How many faces are squares?

170

PUPIL TEXTBOOK 1A PAGE 170

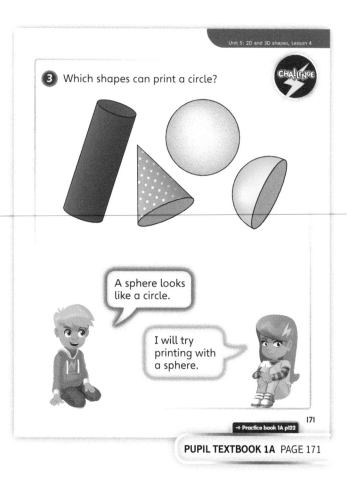

Unit 5: 2D and 3D shapes, Lesson 4

3 Which shapes can print a circle?   CHALLENGE

A sphere looks like a circle.

I will try printing with a sphere.

171

→ Practice book 1A p122

PUPIL TEXTBOOK 1A PAGE 171

Practice

WAYS OF WORKING Pair work

IN FOCUS Question ❶ asks children to match the 3D shapes to the 2D shapes that they print. Question ❷ asks children to cross out the print that 3D shapes cannot make. Question ❸ asks children to identify a 3D shape by the number of faces that the shape has.

STRENGTHEN Children could use concrete 3D shapes to create their own simpler version of the picture in question ❺ and challenge their partner to identify the order in which the shapes were printed. Their partner could also identify the 3D shapes used to print each 2D shape by name.

DEEPEN In question ❷, the rectangle could be made by printing two overlapping squares. Ask children whether there is any way they could use the cube to print the rectangle. Children could try drawing around a cube.

THINK DIFFERENTLY Question ❹ requires children to identify individual 2D shapes from compound shapes. There is often more than one possible answer. For example, the first shape could have been printed using a rectangle and a triangle or a rectangle and a square.

Extend question ❹ by asking children whether there is only one possible solution for each print or whether other shapes could be used.

ASSESSMENT CHECKPOINT Question ❶ will help you decide whether children can identify 2D faces of 3D shapes.

Question ❸ will help you assess whether children can apply their knowledge of 3D shapes to count faces that are not visible.

In question ❹, assess whether children are able to determine basic 2D shapes from composite shapes. In question ❺, check whether children can apply their knowledge of 2D shapes to identify partially obscured shapes.

ANSWERS Answers for the **Practice** part of the lesson can be found in the *Power Maths* online subscription.

Reflect

WAYS OF WORKING Pair work

IN FOCUS This **Reflect** part of the lesson requires children to distinguish between 2D and 3D shapes and match them to their corresponding names.

ASSESSMENT CHECKPOINT Assess whether children are secure in their understanding of which shapes are 2D and which shapes are 3D. Can children label 2D and 3D shapes correctly?

ANSWERS Answers for the **Reflect** part of the lesson can be found in the *Power Maths* online subscription.

After the lesson

- Were any children still using 2D names for 3D shapes or 3D names for 2D shapes?
- Did children have a good understanding of what faces are and the types of faces that different 3D shapes have?
- Were children aware that different 3D shapes may share common shaped faces?

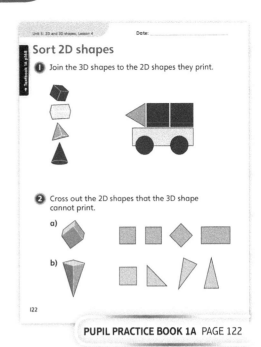

PUPIL PRACTICE BOOK 1A PAGE 122

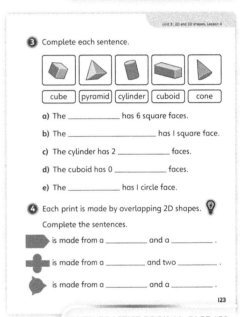

PUPIL PRACTICE BOOK 1A PAGE 123

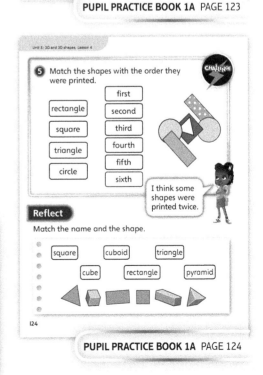

PUPIL PRACTICE BOOK 1A PAGE 124

Make patterns with shapes

Learning focus

In this lesson, children will apply their previous knowledge to identify 2D and 3D shapes within repeating patterns. Children will reason to describe the patterns and to help them identify missing shapes.

Before you teach

- Are children secure naming 2D and 3D shapes?
- Are children able to recognise shapes in different orientations and understand that rotating the shape does not change the shape?

NATIONAL CURRICULUM LINKS

Year 1 Geometry – properties of shape

Recognise and name common 2D and 3D shapes, including: 2D shapes [for example, rectangles (including squares), circles and triangles]; 3D shapes [for example, cuboids (including cubes), pyramids and spheres].

ASSESSING MASTERY

Children can identify and describe the repeated part (the core) of a repeating pattern and reason about how to complete or continue repeating patterns. They can devise their own repeating patterns and summarise the cores of these patterns.

COMMON MISCONCEPTIONS

When trying to find the missing part of a pattern, children may focus only on the missing section and fail to look before or beyond to see how the pattern continues. This can mean that children do not gather enough information in order to complete the pattern. Ask:

- *How do you know what the missing shape is?*
- *Which parts of the pattern did you look at to help you?*

STRENGTHENING UNDERSTANDING

Give children concrete 2D and 3D shapes to print with or draw around. Ask children to replicate the patterns in the questions in this lesson and identify what shapes are being repeated.

GOING DEEPER

Encourage children to create their own repeating patterns and write descriptions to go with them. They can begin to make their patterns more complex by including longer cores and by making more than one line.

KEY LANGUAGE

In lesson: shape, under, rectangle, **pattern**, **repeated**, big, small, triangle, continue, hidden, square, circle, cube, cuboid, sphere

Other language to be used by the teacher: 3D shape, 2D shape, rotate, core

STRUCTURES AND REPRESENTATIONS

3D and 2D representations of cubes, cuboids, spheres, cones, squares, rectangles, triangles, circles

RESOURCES

Mandatory: concrete representations of 2D and 3D shapes: circles (big and small), rectangles, triangles (big and small), squares, cubes, spheres, cuboids, cones, cylinders

Optional: paint and paper for printing

 In the eTextbook of this lesson, you will find interactive links to a selection of teaching tools.

Quick recap

Show children a cylinder. Ask them what 2D shapes they can see on the cylinder. Repeat with a cuboid and a square-based pyramid.

Discover

Make patterns with shapes

WAYS OF WORKING Pair work

ASK

- Questions ① a) and ① b): *Can you describe the patterns on the invite?*
- Question ① a): *How are the circles different?*
- Question ① b): *How do the rectangle and triangle change in the patterns on the invite?*
- Questions ① a) and ① b): *What shape would come before the first shape in each pattern on the invite? How do you know?*

IN FOCUS In question ① b), children identify the pattern and then reason about how many shapes are covered. Children must look beyond the visible shapes to determine the shapes that are covered, and then reason about how many rectangles will fit under the pencil case or mug.

PRACTICAL TIPS Children can use real-life 2D shapes to recreate the patterns in the classroom.

ANSWERS

Question ① a): There is 1 small circle under the mug.

Question ① b): There are 8 rectangles on the right-hand side of the invite.

Discover

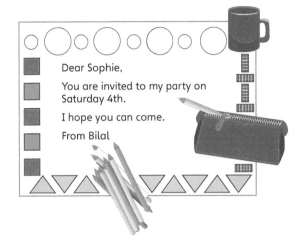

① a) What shape might be hidden under the mug?

   b) How many rectangles are on the invite?

172

PUPIL TEXTBOOK 1A PAGE 172

Share

WAYS OF WORKING Whole class teacher led

ASK

- Question ① a): *How did you know which shape was under the mug?*
- Question ① a): *What would be the next three shapes in the pattern?*
- Question ① b): *How did you work out that three rectangles were covered?*
- Question ① b): *Can you draw the part of the pattern that is covered?*

IN FOCUS The picture in question ① b) provides scaffolding for the reasoning behind solving the problem. Where did children start looking? Starting from the top of the pattern gives more information about the pattern up to the covered shapes. Encourage children to explain how the transparent version of the pencil case helps them to see how many rectangles are covered.

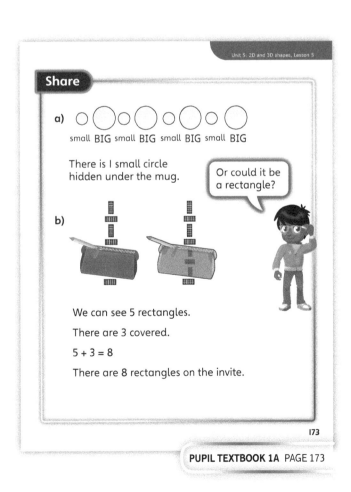

Share

a) small BIG small BIG small BIG small BIG

There is 1 small circle hidden under the mug.

Or could it be a rectangle?

b) We can see 5 rectangles.

There are 3 covered.

$5 + 3 = 8$

There are 8 rectangles on the invite.

173

PUPIL TEXTBOOK 1A PAGE 173

Think together

WAYS OF WORKING Whole class teacher led (I do, We do, You do)

ASK

- *How would you describe the patterns? Which shape would come before the first shape in each pattern? How many shapes are in the repeating part of each pattern? How did you work out which shapes were missing?*

IN FOCUS Question ③ introduces children to a pattern that has three shapes in its core, illustrating that patterns are not limited to two repeating elements. This means that children need to look at more of the pattern in order to determine the hidden shape.

STRENGTHEN Have concrete 2D and 3D shapes available so children can recreate each pattern and identify the core by separating it from the rest of the pattern. Cover one or more shapes and ask: *Which shape is covered?* Children could continue this with a partner.

DEEPEN Encourage children to apply their knowledge of each pattern beyond what is represented on the page. For example, in question ② a), ask children to predict the seventh shape in the pattern and justify their choice.

ASSESSMENT CHECKPOINT In question ①, assess whether children are able to identify which part of the pattern is being repeated.

In questions ② and ③, assess whether children can describe the patterns to their partner. Prompt children by referring them to the language used by Dexter and Astrid, especially the words 'repeated' and '3D'.

ANSWERS

Question ①: It is ▽△ because the size of the triangles and the direction they face matches the pattern.

Question ② a): ◇

Question ② b): ▪

Question ③: ▢ The pattern repeats cube, sphere, cuboid.

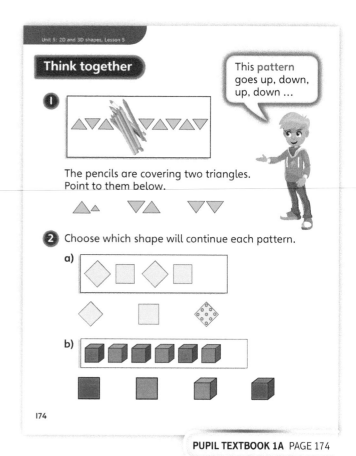

PUPIL TEXTBOOK 1A PAGE 174

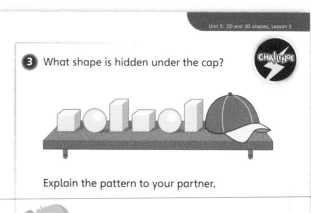

PUPIL TEXTBOOK 1A PAGE 175

Practice

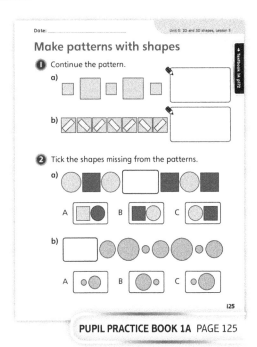

PUPIL PRACTICE BOOK 1A PAGE 125

WAYS OF WORKING Pair work

IN FOCUS Questions ❶ and ❷ a) show repeating patterns with two shapes in the pattern. Questions ❷ b), ❸ and ❹ move on to show patterns with more than two shapes in the pattern.

Question ❺ uses a pattern that is visually very different. Children need to calculate the missing number in the part-whole model, but the pattern is determined by its orientation. Discuss where the numbers need to go and how the shape has changed: has it been flipped or rotated?

STRENGTHEN Have concrete 2D and 3D shapes for children to use in order to replicate the patterns. Encourage them to point out and name each shape to their partner. Ask: *Can you hear the repeating part of the pattern?*

DEEPEN Ask children to predict the 10th, 11th and 12th shapes in each pattern. How do they know what these shapes will be?

THINK DIFFERENTLY Question ❸ asks children to circle the repeating pattern and write the number of shapes in the repeating pattern. Question ❸ b) uses a cone twice in the core, which means that identifying the core becomes more complex. Encourage children to look closely and to name the shapes as they go down the line.

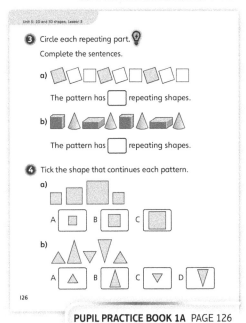

PUPIL PRACTICE BOOK 1A PAGE 126

ASSESSMENT CHECKPOINT In question ❷, determine whether children can apply their knowledge of a pattern to work out which shapes are missing. Do children look at enough of the pattern in order to identify the repeating part and the missing shapes?

Question ❸ will expose whether children are able to identify and describe the core of the pattern. Are children looking at enough of the pattern in order to identify the core?

ANSWERS Answers for the **Practice** part of the lesson can be found in the *Power Maths* online subscription.

Reflect

WAYS OF WORKING Pair work

IN FOCUS The **Reflect** part of the lesson requires children to apply their knowledge of patterns to create their own repeating pattern. Encourage children to challenge their thinking by using one or more of the shapes more than once in their pattern's core.

ASSESSMENT CHECKPOINT Assess whether children can describe their pattern's core. Have they continued the pattern long enough so the core can be identified? The complexity of their patterns can help determine how secure children are in their understanding of repeating patterns.

ANSWERS Answers for the **Reflect** part of the lesson can be found in the *Power Maths* online subscription.

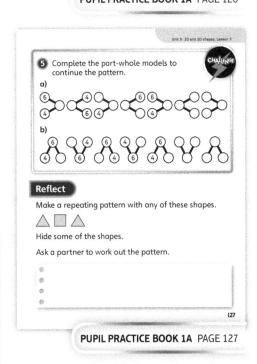

PUPIL PRACTICE BOOK 1A PAGE 127

After the lesson

- Did children understand that a pattern needs to have a core?
- How confident were children in identifying and describing the core of a pattern?
- Did children develop an understanding of and ability to use cores with three or more elements?

End of unit check

> **Don't forget the unit assessment grid in your *Power Maths* online subscription.**

WAYS OF WORKING Group work adult led

IN FOCUS

Questions ❶ and ❷ assess children's recognition of 3D shapes, including when they are presented in different orientations.

Question ❸ assesses children's recognition of 2D shapes, in particular triangles, including when they are presented in different orientations.

Question ❹ assesses children's ability to recognise patterns using 2D shapes and their written names.

Question ❺ focuses on vocabulary and distinguishing between 2D and 3D shapes.

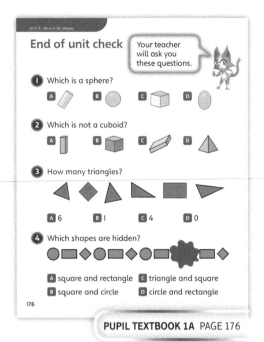

PUPIL TEXTBOOK 1A PAGE 176

Think!

WAYS OF WORKING Pair work or small groups

IN FOCUS

- This question assesses children's ability to distinguish between 2D and 3D shapes.
- Draw children's attention to the words at the bottom of the **My journal** page. Which group(s) does each word describe?
- Encourage children to think through or discuss how the shapes in each group are the same before writing their answer in **My journal**. Can they match the 2D shapes to the 3D shapes?

ANSWERS AND COMMENTARY Children will be able to describe the differences between 2D and 3D shapes using the correct mathematical vocabulary. They will be able to name and describe the properties of a variety of 2D and 3D shapes.

THINK QUESTION Children should indicate that the shape belongs in the first group as it is a 3D shape, not a 2D shape.

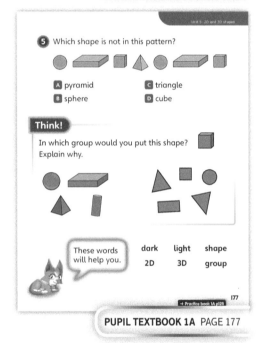

PUPIL TEXTBOOK 1A PAGE 177

Q	A	WRONG ANSWERS AND MISCONCEPTIONS	STRENGTHENING UNDERSTANDING
1	B	A or D suggests that the child has identified the curved face and associated this with what they know about a sphere.	Provide children with 2D and 3D shapes to manipulate and explore. Ask them to match the shapes with those in the problems. If they are still having difficulties correctly naming the shapes, provide labels and ask them to match them with the shapes, talking about their properties in the process.
2	D	A suggests that the child does not recognise a cuboid with 'unusual' dimensions, despite its familiar orientation.	
3	C	B suggests that the child only recognises regular triangles.	
4	B	A, C, or D suggest that the child could not identify the pattern's core or that they do not understand the shape vocabulary.	Ask children to sort the shapes and discuss the criteria that they have used. They could be prompted to sort them by their properties; for example, curved surface / not curved surface or six faces / not six faces.
5	C	A suggests that the child has misidentified the pyramids pictured as triangles.	

My journal

WAYS OF WORKING Independent thinking

ANSWERS AND COMMENTARY

The cube belongs in the group of 3D shapes.

Possible explanations include:
- I put the shape there because it is a 3D shape like the rest in the group.
- I put the shape there because it has a dark face like the rest in the group.
- I put the shape there because it is shaded like the rest in the group.
- I put the shape there because the other group has 2D shapes.

If children are unable to correctly place the shape or justify their choice appropriately, they may need more time grouping and sorting 2D and 3D shapes practically. Ensure that you encourage children to give a running commentary, to promote mathematical language and clear articulation of their thinking.

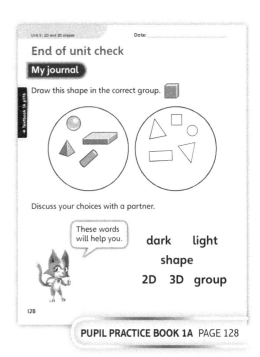

PUPIL PRACTICE BOOK 1A PAGE 128

Power check

WAYS OF WORKING Independent thinking

ASK
- *How do you feel about naming and describing 2D and 3D shapes?*
- *Were there any parts that you found challenging? Why?*

Power puzzle

WAYS OF WORKING Pair work or small groups

IN FOCUS Use this **Power puzzle** to see if children can use the names of 2D shapes when discussing the problem with a partner. Attempt the puzzle in front of the class and break the rules to see if children understand them and can identify where you went wrong; for example, use more than three colours or colour two adjacent shapes the same colour. When children are secure, ask them to take turns colouring while their partner gives advice about what to do. This will encourage children to be clear in their descriptions and use correct mathematical vocabulary.

ANSWERS AND COMMENTARY Children who complete the **Power puzzle** successfully, giving clear descriptions of the shapes as they do so, can identify shapes based on their properties and have a good understanding of pattern. If children cannot give their partner clear advice, or cannot follow their partner's advice, they may not be secure with identifying and naming 2D shapes.

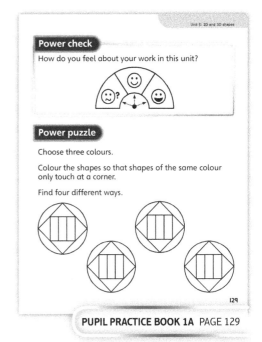

PUPIL PRACTICE BOOK 1A PAGE 129

After the unit ⏸

- How can you find opportunities to reinforce the learning of this unit by identifying shapes in everyday contexts?
- Can you make cross-curricular links, drawing on children's learning from this unit?

Strengthen and **Deepen** activities for this unit can be found in the *Power Maths* online subscription.

Published by Pearson Education Limited, 80 Strand, London, WC2R 0RL.

www.pearsonschools.co.uk

Text © Pearson Education Limited 2018, 2022
Edited by Pearson and Florence Production Ltd
First edition edited by Pearson, Little Grey Cells Publishing Services and Haremi Ltd
Designed and typeset by Pearson and Florence Production Ltd
First edition designed and typeset by Kamae Design
Original illustrations © Pearson Education Limited 2018, 2022
Illustrated by Nadene Naude, Kamae, Daniel Limon, Adam Linley, Laura Arias, Jim Peacock,
Eric Smith, Nigel Dobbyn and Phil Corbett at Beehive Illustration, Kamae Design, and Florence
Production Ltd
Cover design by Pearson Education Ltd
Back cover illustration © Will Overton at Advocate Art and Nadene Naude at Beehive Illustration
Series editor: Tony Staneff; Lead author: Josh Lury
Authors (first edition): Tony Staneff, David Board, Natasha Dolling, Caroline Hamilton,
Julia Hayes and Timothy Weal
Consultants (first edition): Professor Jian Liu

The rights of Tony Staneff and Josh Lury to be identified as authors of this work have been
asserted by them in accordance with the Copyright, Designs and Patents Act 1988.

This publication is protected by copyright, and permission should be obtained from the
publisher prior to any prohibited reproduction, storage in a retrieval system, or transmission
in any form or by any means, electronic, mechanical, photocopying, recording, or otherwise.
For information regarding permissions, request forms and the appropriate contacts, please
visit https://www.pearson.com/us/contact-us/permissions.html Pearson Education Limited
Rights and Permissions Department

First published 2018
This edition first published 2022

26 25 24 23 22
10 9 8 7 6 5 4 3 2 1

British Library Cataloguing in Publication Data
A catalogue record for this book is available from the British Library

ISBN 978 1 292 45049 0

Copyright notice
All rights reserved. No part of this publication may be reproduced in any form or by any means
(including photocopying or storing it in any medium by electronic means and whether or not
transiently or incidentally to some other use of this publication) without the written permission of
the copyright owner, except in accordance with the provisions of the Copyright, Designs and Patents
Act 1988 or under the terms of a licence issued by the Copyright Licensing Agency, Barnards Inn,
86 Fetter Lane, London EC4A 1EN (http://www.cla.co.uk). Applications for the copyright owner's
written permission should be addressed to the publisher.

Printed in the UK by Ashford Press Ltd

For Power Maths online resources, go to:
www.activelearnprimary.co.uk

Note from the publisher
Pearson has robust editorial processes, including answer and fact checks, to ensure the
accuracy of the content in this publication, and every effort is made to ensure this publication
is free of errors. We are, however, only human, and occasionally errors do occur. Pearson is
not liable for any misunderstandings that arise as a result of errors in this publication, but it is
our priority to ensure that the content is accurate. If you spot an error, please do contact us at
resourcescorrections@pearson.com so we can make sure it is corrected.